The Beginner's Handbook

The
CLOCK
REPAIR
PRIMER

Philip E. Balcomb
Fellow, NAWCC

A Beginner's Introduction

to

THE MECHANICS OF PENDULUM CLOCKS

and

Basic CLOCK REPAIR

With

144 Illustrations by the Author

TEMPUS PRESS, Tell City, Indiana

DEDICATION

This book is dedicated to one who strongly resisted my early clock collecting involvement, with its attendant drain on my time and our limited resources, but who came to be my most enthusiastic and dedicated supporter in this activity. Absent her patience and forbearance during the two years this book has been in preparation, it would never have become reality. -My wife, helpmate and dear friend of forty-five years, Virginia.

PUBLISHING HISTORY

This is a revised and expanded version of The Clock Repair Primer which was published in serial form in the NAWCC BULLETIN, official publication of the National Association of Watch and Clock Collectors, Inc., in seven successive issues, beginning with the February, 1985 issue and completed in the April, 1986 issue.

Library of Congress Catalog Card No. 86-051000

Balcomb, Philip E.

        The Clock Repair Primer
        The Beginners Handbook

        The Mechanics of Pendulum Clocks and
            Basic Clock Repair
        Bibliography
        Index
        Source Reference

Published by Tempus Press
PO Box 235, 104 Geneva Drive
Tell City, IN 47586

SIXTH EDITION - January, 1995

ISBN 0-9620456-0-8

Manufactured in the United States of America

TABLE OF CONTENTS

AUTHOR'S FOREWORD

I have been deeply touched by the many complimentary letters and comments made to me and the constructive comments and suggestions I have received from fellow members of NAWCC, since publication of this work in serial form started in the NAWCC BULLETIN. Many of their ideas have been incorporated in this revised version.

For almost two years this project has been my primary creative concern. It has led me to question a great many details I had formerly considered of little concern and demanded careful analysis and research of a number of subjects. This has been an enlightening and rewarding experience.

A significant debt of gratitude is owed, and acknowledged, to James W. Gibbs, who encouraged, gently prodded, and helped formulate the basic concept of this effort. Terry Casey, Editor of the NAWCC BULLETIN has been helpful beyond measure, in areas both technical and editorial. Last, but not least, I thank all of the many members of NAWCC who have helped me learn about clock repair and countless other horological subjects, greatly enriching many years of my life.

I particularly acknowledge the personal help given me in my earliest years of interest in clocks by two members of Indiana Chapter 18, Tom Ayton, long time Chapter President and Al Osborne, author of a series of articles on clock repair, "Doing it the Osborne Way",

published in the NAWCC BULLETIN many years ago. They were always ready to respond to what I thought might be stupid questions with carefully considered answers.

It is my sincere hope that this book will help future generations of collectors obtain an early understanding of clock mechanisms and give them an easier start than I had in the fundamentals that add so much to the fascinating hobby of clock collecting.

Only yesterday, while in the process of final preparation for the printer, I received my copy of the August, 1986 NAWCC BULLETIN. In the Vox Temporis (Voice of Time) section I came across a letter containing a paragraph that brought a lump to my throat. Charles K. Aked of West Drayton, Middlesex, England, wrote:

"And in closing, a word of congratulation to Philip E. Balcomb. There will be many whose future knowledge will rest on the basis of what he is writing today. Too many experienced horologists forget that we all have to make a start somewhere and amongst the beginners of today are some who will be the horological giants of tomorrow"

Should this little book for beginners be the spark that lights even one small fire to add warmth and joy to life and the appreciation of horology, I shall consider my time well spent.

Philip E. Balcomb
Tell City, Indiana
July 30, 1986

Introduction

This book is not intended for clock makers, nor for the advanced amateur. It has been designed to enable ordinary people to perform simple repair of spring or weight driven pendulum clock movements. It is truly a Primer (rhymes with dimmer), written in layman's terms, in a simple step-by-step, manner.

It is hoped that it may lead to a better understanding of the parts that make up a clock movement, and how they function together. Increased awareness of the basic elements of movements can add another dimension to the enjoyment of collecting clocks. Nothing can match the thrill of seeing a once inanimate clock restored to life, as a result of one's own efforts.

A number of books on the repairing of clocks are available which can provide useful additional information on specific subjects. A bibliography listing some of these will be found at the end of this book. For the novice, most such publications pose a problem. They assume a basic knowledge of clock mechanisms and, in many cases, presume the reader has available relatively sophisticated tools and equipment. Generally, it is difficult to locate specific subject matter and they use horological terms, not readily understood by the novice.

Fortunately, many repairs can be made safely and effectively by anyone with reasonable manual skills, a strong desire to learn and the patience to study and follow instructions. We must emphasize, however, that many clock movements are relatively delicate and must be treated gently, to avoid serious damage.

USING THIS PRIMER

In this work, we will assume the reader is totally unfamiliar with how clocks work, why they stop and what can be done to restore them to running order. We will discuss only procedures that can be readily accomplished with ordinary hand tools or with the aid of inexpensive or home made tools and equipment. You will be led through the movement in simple, logical steps.

A subject Index is provided, at the back of the book, to help you locate various aspects of a subject referred to in different places in the text.

WORKING WITH A MOVEMENT AT HAND

You will find that the step by step logic of this book will be most understandable if you have a clock movement in hand as you go through it. While we have included a number of illustrations of some of the critical elements, there can be no substitute for having the real thing available for observation and comparison with the text and drawings.

Follow each step, as it applies to your clock, including taking the movement out of the case. This will provide an interesting experience and will give you a good understanding of how clocks work. As you will see, repair of problems that can cause clocks to stop running is often quite simple.

How far you go with repairs will depend on you. If you take it a step at a time, you may be surprised at how much you will be able to do, with confidence.

HOW THIS PRIMER IS ORGANIZED

The CLOCK REPAIR PRIMER is unique. It is the first book in the entire history of clock making, to begin at the beginning and show the reader in logical sequence how a clock works and the many simple repairs he can make. You will be led through the movement, step by step, beginning with the first moving parts visible to the casual observer, the hands. We will explain many, if not all, of the variations you can expect, along with simple instructions for correcting deficiencies. Simple, but thorough, illustrations will even enable you to accomplish disassembly, cleaning, simple repair, lubrication and reassembly.

In addition to being arranged in a logical sequence, individual subjects are captioned and indexed so that they can be easily found. Horological (clock makers) terms, where used, are explained.

CAUTION

Some procedures, let's face it, are beyond the capability of all but the professional, or advanced amateur. You will be helped to recognize the current limit of your skills so that you can avoid inflicting severe injury on a valued clock. You are urged to learn basic principles and procedures by practicing on the simpler movements recommended.

Practicing on simple, readily available, and relatively inexpensive movements, as we will suggest, will enable you to learn, without worrying about possible damage to a valued clock.

WHERE TO FIND HELP

When you run into a problem and need help, you can frequently find it among the members at Chapter meetings of the National Association of Watch and Clock Collectors, one of which is located in your area.

Membership is open to anyone interested in clocks or watches. On alternate months, NAWCC publishes and distributes to its members its BULLETIN, a magazine which contains articles on all aspects of clocks and watches. In months between those in which the Bulletin is published, the Association sends the MART to all members. This publication regularly carries ads by members capable and willing to do advanced repair work for you, as well as clocks and watches available for sale or trade.

For further information, write

NAWCC,
P.O. Box 33, PB
Columbia, PA 17512.

IN CONCLUSION

The material in this book is the result of many years of spare time study and experience, by an ordinary amateur repairman-collector, and is intended for other hobbyist-collectors. It is meant to supplement, not replace, more advanced clock repair books, or the invaluable services of skilled clock makers.

The author has diligently tried to eliminate error and has included only procedures he has found practical in his personal experience. Comments and suggestions for improvement are solicited and will be most welcome. In using this book, the reader is on his own and no liability of any kind shall accrue to the author or publisher.

Chapter 1

CLOCK HANDS

The hands are the ultimate moving parts of the clock, although we seldom think of them as a part of the movement (works) and possibly responsible for malfunction of the clock. They are the last parts attached to the movement and, since they are in front of the dial, must usually be removed before the movement can be inspected.

I. HAND MALFUNCTIONS

When a clock refuses to run, the hands may be the cause. In many cases, they are of relatively flimsy construction and can easily be bent during setting. Since the minute hand is on top of the hour hand, if it is bent toward the face it may come into contact with the hour hand and be unable to get past it, stopping the clock. Because the minute hand is longer than the hour hand, it is possible for its tip to be bent down, so that it rubs on the dial, causing enough friction to stop the clock.

If the clock has stopped with the hands adjacent to each other, it is likely they have interfered with each other.

If the hands are not together and the clock refuses to run, before proceeding further, check to see if the tip of the minute hand is in contact with the dial face. Look also to see if there are marks on the dial that would indicate the tip of the hand has been dragging. In the event the minute hand has been bent significantly away from the dial, it may rub against the dial glass. If any of these conditions are present, you may have found the reason for stoppage.

B. REPAIR PROCEDURE:

1. Hands in contact:

Bearing in mind that some hands may be extremely delicate and easily broken, move the minute hand clear of the hour hand and grasp it, near its base, then gently bend it away from the face. After doing so, carefully move it forward (never backward) to the hour hand and see if it clears comfortably. If not, repeat the bending operation and try again. It is better to do this in easy stages than to bend too much at one time.

The clearance between the hands should be such that they are about parallel. That is, the space between them from their points of attachment to the movement to the end of the hour hand should be uniform. If the hand has been bent too far out, press near its hub and bend it back, until proper spacing is attained.

If the minute hand rubs on the glass, gently press on it, near its hub, until it is parallel to the face and hour hand.

II. BROKEN HANDS:

Delicately made hands, such as those found on many French clocks, are brittle and easily broken if too much force is used.

A. REPAIR PROCEDURE:

If you have the parts, an experienced repairmen may be able to solder them together for you, using hard silver solder. In order to provide sufficient strength, he may have to add a tiny reinforcing piece on the back side, across the break.

Unless you are proficient in silver-soldering, you should seek help. NEVER USE SOFT SOLDER for any clock hand or movement repairs. It is not strong enough and is bound to fail.

B. REPLACEMENT:

Suitable replacement hands are generally available. See the catalogs of clock parts suppliers, some of which are listed in the Appendix. Many hand styles are illustrated. Choose the style and size that most closely matches the broken ones.

III. BASIC TYPES OF HANDS:

Clock hands can be found in literally thousands of shapes and sizes, the oldest ones hand made by highly skilled craftsmen. For our purposes, we will ignore the outer part of the hand and deal only with the way it is attached to the movement. There are a limited number of variations in the basic shapes of the holes in the hubs commonly used for this attachment; round, square, or oblong. There are two basic types of hands:

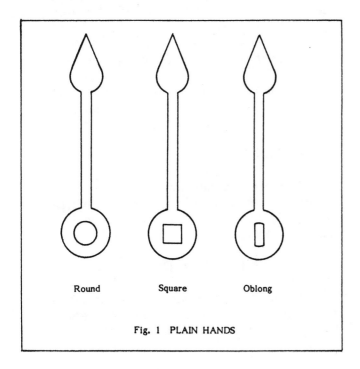

Round   Square   Oblong

Fig. 1 PLAIN HANDS

A. PLAIN HANDS:

Hands made in one piece, usually of steel, with a hole in the hub, are called "plain".

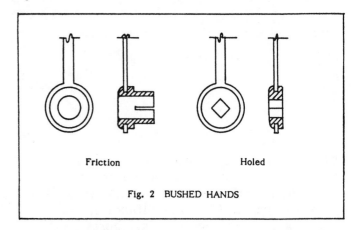

Friction   Holed

Fig. 2 BUSHED HANDS

B. BUSHED HANDS:

Hands which are otherwise like plain hands, but have a tightly fitting brass bushing, with a hole in it, attached to the hub, are known as "bushed" hands.

IV. CLOCK HAND FUNCTION:

When you have determined that the hands are not what is causing the clock not to run, you must look further. Before the movement can be inspected, it will be necessary to remove the hands and then the dial.

At this point, you should understand something about the way the hands are attached and driven and how they function. Figure 3 shows the parts of the movement that support both minute and hour hands.

A. The MINUTE HAND:

This hand is attached to a solid shaft, called an "arbor". This arbor must be able to turn, independently of the other arbors in the time train, when setting the minute hand to the correct time. Between the arbor and the pinion which drives it, is an eared spring washer or other device that acts as a clutch, as we will see later.

B. The HOUR HAND:

The hour hand is attached to a hollow

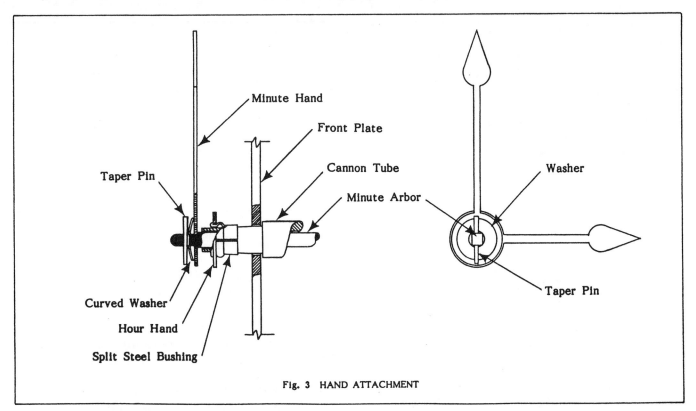

Fig. 3 HAND ATTACHMENT

tube which fits over the minute hand arbor. This tube, which is referred to in horological (a fancy word meaning "related to time keeping") jargon as the "cannon", makes one revolution in 12 hours.

V: HAND ATTACHMENT and REMOVAL:

Not only do hands vary in their size and shape, they also evidence considerable variation in the way they are attached to the minute arbor and cannon tube. Before attempting to remove them, it is important to determine what method of attachment has been used on the particular clock on which you are working.

A. MINUTE HANDS:

Minute hands of striking clocks must be securely attached to the minute arbor in a fixed position, so they are always point to the 12 or 6 on the dial when the clock strikes. To insure this, they nearly always have a square or oblong hole designed to closely fit a similar section on the end of the arbor. They are held in place by various means, as we shall see. Time pieces (non-striking clocks) also use this system simply to insure that the hand does not slip on the arbor when using the hand to set the time.

1. Taper Pin Retained Hands:

A taper pin is a short piece of steel or brass which is smaller at one end than at the other. It passes through a small hole in the end of the minute shaft. When pushed firmly into place, it is held there by friction. Usually, there are one or more semi-spherical washers between the pin and the hand, so that the pin exerts pressure on the washers to firmly secure the hand.

All too frequently, the taper pin has been replaced by a small nail or straight pin. A new taper pin should be substituted.

a. Removing the Taper Pin:

Large movements will generally have larger minute hand shafts than smaller movements. Hand holes, washers and taper pins will also be larger. The smaller the taper pin, the more subject it is to damage during removal. Care should be exercised in removing any taper pin, to avoid possible damage to the hands or dial.

Perhaps the most common method of removal is the use of pliers, either to grip the large end of the pin and pull it out of the hole, or by placing one jaw against the shaft and the other on the small end of the pin, forcing the pin out of the hole. While these methods work, at the very least, scoring of the pin is likely if the pull method is used and the pin may be bent in the push method. Fortunately, replacement pins are available.

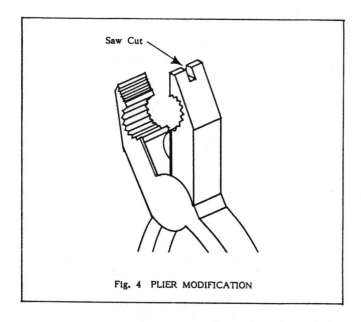

Fig. 4 PLIER MODIFICATION

(1) Making a Pin Removing-Replacing Tool:

An inexpensive taper pin removing tool can easily be made from a cheap pair of 6" flat-nosed (lineman's) pliers, without affecting their overall utility. Open the pliers to their widest position and hold them in a vise so that the end of one jaw is exposed about 1/4" above the vise. With a hacksaw, make two cuts, side by side at the center of the jaw and about 1/8" deep, to form a groove. That's all there is to it.

(a) Using the Modified Pliers:

Place the groove in the jaw of the pliers over the large end of the taper pin and against the side of the minute arbor, with the other end bearing on the small end of the pin and squeeze. The pin will come out easily and without bending, because the pressure was exerted on a straight line. Reversing the position of the jaws makes re-insertion of the pin just as easy.

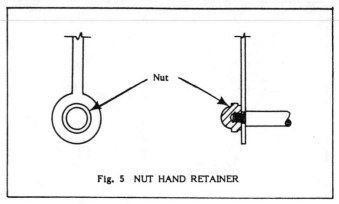

Fig. 5 NUT HAND RETAINER

2. Threaded Nut Retained Hands:

This method of attachment of the minute hand is easily recognized by the presence of a nut on the end of the arbor. If the nut is missing, the exposed end of the arbor will be threaded and there will be no taper pin hole. Hand nuts are found in many shapes, from hex to plain round, to round domed. There is sometimes, but not always, a flat or semi-spherical washer between the nut and the hand.

a. Removing the Nut

The hand nut is frequently only finger tight and you should try unscrewing it with your fingers. If this is not successful and the nut is hex shaped, use a small open-end or crescent wrench, holding the hand to offset the pressure of turning the nut.

If the nut is round, use flat or long nosed pliers and gentle pressure while turning. To avoid marring the nut, or the minute hand, you may want to use a piece of soft cloth between the plier jaws and the nut. As soon as the nut turns easily, finish removing it with your fingers.

3. Washer Retained Hands:

A few clocks employ a round, sometimes slightly spherical, washer, with a square hole, to retain the hands. The absence of a pin or a retaining nut and the presence of a fairly large steel washer, on top of and firmly in contact with the hand, indicates this system of attachment.

In this method, the square end of the

shaft has a shallow groove, a short distance from the end. A square hole in the washer fits closely over the square end of the shaft. When the washer is rotated at the level of the groove in the shaft, the flat sides of its square hole enter the groove and are retained by the corners of the shaft square. Friction holds it in place.

a. Removing a Square Holed Retaining Washer:

This is probably the most difficult type of hand retainer to remove. Try pressing against the hub of the hand with the fingers of one hand, compressing the clutch washer on the arbor, while turning the washer with those of the other. If this is not successful, it will be necessary to use pliers with good, square jaw edges.

Placing the sides of the plier jaws flat against the face of the hand, grip the sides of the washer and turn it. Be very careful in doing this, to avoid marring of the washer or hand.

4. Malfunctions of Minute Hands:

One of the most common complaints of clock owners is that the minute hand is not near the figure 12 on the dial, when striking occurs. While more common with older clocks, this has been known to happen with new clocks. We refer to this condition as being out of synchronization with the strike train of the movement. It usually does not indicate internal problems.

If, however, the hand is firmly secured to the arbor and yet can be moved back and forth, a minute or two, without meeting any resistance, the cause is probably accumulated wear of the teeth of wheels and pinions in the movement. When this is the case, gravity will cause the hand to advance slightly after it passes 12, and to slip back the same amount as it passes 6. Such an error usually is less than two minutes and there is no practical cure.

Minute hands can be out of synchronization because of wear of the hole in the

hand, or wear of the part of the arbor on which they fit. This is more likely to be true of plain hands, but may also happen with bushed hands.

Another cause of lack of synchronization, with bushed hands, is that the bushing may have become loose enough to allow slight rotation, or was not firmly secured in the hand at the time of manufacture.

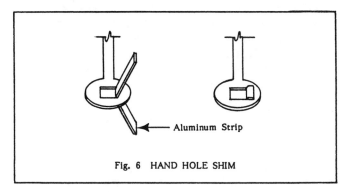

Fig. 6 HAND HOLE SHIM

1. Correcting Plain Hand Synchronization:

Since wear is the cause of this problem, when plain hands are involved, material has been worn away from either the hole in the hand, or the end of the arbor. Only slight wear of either or both can result in the hand position being a few minutes off.

This is a very common problem with old wood movement clocks, which have a wood minute arbor, easily damaged when the metal minute hand is removed and replaced. The method of correction is the same for wood and brass movements.

For many years this problem has been corrected by the best repairmen in a way that is still probably the best. It consists of adding material to the sides of the hand hole, to reduce its dimensions and provide a tighter fit on the arbor.

Traditionally, very thin sheet zinc was used. Since it is not readily available today, we have found a good substitute in the thin, quick opening, aluminum lids found on containers for nuts and snack foods.

a. Procedure for Correcting:

Using sharp sheet metal shears, or kitchen shears cut a strip of aluminum very slightly narrower than the width of the hole in the hand. Insert the strip in the hand hole, against the side opposite the hand and bend both ends flat against the side of the hole.

Place the hand on the arbor. It should fit much more snugly than before. If it is not reasonably snug, add another strip of aluminum at the opposite side of the hole. Rotate the hand slowly, until striking occurs. If the hand points to 3, 6 or 9 at the strike,, remove it and replace so it points to 12. Again rotate the hand, very slowly as you near 12 and observe how close it is to 12 when the clock strikes. If it is within a minute or two, you have cured the problem. Secure the hand to the arbor with washer and taper pin, or nut.

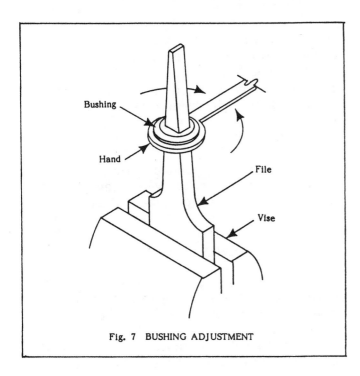

Fig. 7 BUSHING ADJUSTMENT

2. Correcting Bushed Minute Hand
    Synchronization:

This correction involves changing the position of the hole in the bushing, with respect to the arm of the hand. While the bushing may have moved slightly, it is probably still relatively tight to the hand and some force will be required.

a. Procedure for Correction:

With the hand in place on the arbor, slowly rotate it to the point where the clock strikes. Carefully note how many minutes it is in error, to the left or right of 12.

Find a small file whose tang will fit well in the square hole of the bushing. Secure the file, tang end up, in your vise, then place the hole in the bushing snugly on it.

Carefully grip the hub of the hand with a pair of pliers and, using minimum pressure, move the hand the desired amount. It is better to underestimate the needed correction and, after checking it again on the clock, repeating the process until striking occurs at 12.

B. HOUR HANDS:

Hour hands fit on the cannon tube and are of two basic types. One type is used commonly on movements without strike trains (Timepieces), or those with count wheel strike systems, where the position of the hour hand does not have to relate to the mechanics of striking.

The second type is used on movements equipped with rack-and-snail or other strike systems in which the clock is caused to strike a given number of times, every time the cannon rotates to the same position. For example, if the hands are properly in place, the minute hand may be turned past the hour several times without allowing the clock to fully strike, but when the hand is moved to 12 and the clock permitted to strike, it will strike the number of times indicated by the hour hand on the dial.

1. Push Fit Hour Hands:

Used on timepieces or movements with count wheel strike, these hands are attached by simply pressing them onto the cannon. The cannon tube is very slightly tapered, so that the round hole in the hand slips easily over the end of the cannon, but soon meets with resistance. Further pressure causes it to be held securely in place by friction. Since they

6

are friction fit, such hour hands may be moved on the cannon tube to the number on the dial matching the hour struck.

Some plain hour hands have only a round hole, with no bushing, as in Fig.1. Others have a sort of flange created when the hole is punched. Not infrequently, there is a narrow slit, opposite the arm of the hand, to permit some expansion of the hole as it is pressed on the cannon.

When putting plain hour hands on the cannon, it is important to make sure they are parallel to the face of the dial, since there is no bushing to align them.

a. Hand Bushings: (Figs. 2 & 3)

Most hour hands are made of steel with a hole at the hub end into which is inserted a split bushing intended to tightly grip the cannon tube.

(1) Sheet Steel Bushings: (Fig. 3)

Used on many American and other clocks, these bushings are formed of a very thin piece of steel to make a cylinder with a slight gap. Near its outer end is a small shoulder which fits tightly against the back of a hole in the hand. A short section goes through the hole and is bent over and crimped tightly against the face of the hand to hold the bushing securely in place in the hand.

The sides of the bushing are straight. When it is pushed over the tapered cannon tube, the slot allows it to stretch slightly, creating pressure to hold the hand in place.

(2) Brass Hand Bushings: (Fig. 2)

Hands of better quality movements usually have machined brass bushings which perform the same function as sheet steel bushings, but are of sturdier construction. They also have a shoulder that fits tightly against the back of the hand hole and a small projection through the hole that is tightly crimped to secure them to the hand. Only the portion of the bushing below the shoulder is split and the walls are quite a bit thicker than those of sheet steel bushings.

b. Removing Push-Fit Hour Hands:

Usually, it is possible to remove these hands simply by grasping the hub end with the tips of the thumb and second finger and pulling the hand off the cannon tube. Often, however, the fit is so tight that other methods must be used.

CAUTION: Hands can be easily damaged by too much force. Before exerting any force, beyond fingertip pulling, be sure you are dealing with a push-fit hand, not one retained by a nut or washer. When using force, be gentle and stop the instant you detect any bending of the hand.

(1) Rotate the Hand:

Grasp the hand with the fingers, near the hub and try to rotate it on the cannon tube. If it can be moved, without bending the hand, continue rotating back and forth, pulling up on it as you turn. If this is not successful, it will be necessary to use pliers.

(2) Using Pliers:

Pliers used to remove push-fit hour hands should have serrations (small grooves) running crosswise on the insides of the jaws. Carefully position the nose of the pliers perpendicular to the face of the hand and gently grip the sides of the hand, at the center of the hub, with the tip of the jaws. Twist, to rotate the hand very slightly, as you pull the hand from the cannon tube.

This must be done carefully. Too much pressure on the pliers will result in serious bending of the flange of the hand.

c. Malfunctions of Push-Fit Hands:

Since these hands are held in place by friction, both on the cannon tube and where the hand is attached to the bushing, if the metal is stretched, at either location, the hand becomes loose.

(1) Correcting Loose Fit on Cannon Tube:

With both types of bushings, this involves reducing the space in the slits, slightly reducing the inside diameter of the bushing, so that it will fit more tightly on the cannon. Using pliers, gently apply and release pressure on the bushing as you rotate it, to maintain its roundness. Test the fit on the cannon tube. Repeat this procedure until the bushing fits snugly on the tube.

(2) Correcting Loose Fit of Bushing in the Hand:

(a) Sheet Steel Bushing:

Measure the diameter of the outside of the bushing, near the shoulder, and drill a hole of that size in a piece of hard wood. Insert the bushing into the hole so that the shoulder fits snugly against the top of the hole. Use a pencil point, or other tapered instrument to press the bushing as tightly as possible against the edges of the hole in the hand. A light tap with a small hammer will force the bushing outward.

Now, using a flat-nosed punch or a nail whose point has been filed flat, and a small hammer, gently tap around the edge of the bushing that holds it to the hand. Continue tapping progressively around the bushing until it is tight to the hand.

(a) Machined Brass Bushing:

Follow the same procedures as indicated above for sheet steel bushings, but, since brass is much softer than steel, be especially careful not to exert too much force in squeezing and tapping.

2. Square or Oblong-Holed Hour Hands:

Many clocks, particularly those equipped with a rack-and-snail type strike system, have hour hands with square or oblong holes which fit over a mating section on the end of the cannon tube. They are usually held in place by a washer with a suitable hole that fits on top of the hour hand and bears against the back of the minute hand. Frequently, but not always, the hole is contained in a brass bushing secured to the hand in the same manner as those in push-fit hand bushings. (Fig. 3)

The reason for using this type of hole is usually to position the hand accurately on the dial, so that it properly indicates the time on the dial at the instant of activation of rack and snail strike trains. You may, however, encounter them reasonably often on other types of movements, notably fine European and wood movement American clocks with count wheel strike and on some timepieces.

a. Malfunctions of Square or Oblong-Holed Hour Hands:

(1) Looseness on Arbor:

These hands may become loose on the cannon tube when the domed washer that retains them in position under pressure from the minute hand becomes inadequate. In this case, add one or more washers, available from clock supply houses, to increase the pressure.

(2) Wrong Position:

Hands must be positioned on the square of the tube so that they indicate the hour when the clock strikes that hour. To insure this, temporarily place the minute hand on its square and rotate until the clock strikes the hour. Remove the minute hand and place the hour hand on its square, pointing to the hour struck, and secure. Then place and secure the minute hand so it is at the 12 o'clock position.

VI. CLEANING AND RESTORING HANDS:

Hands, like other parts of a clock, can become dirty, or rusted, and should be cleaned. This should be done only after they have been removed from the movement.

A. CLEANING:

Place the hand on a flat surface, face up, with the bushing over the edge, so that the hand is fully supported. Using a soft cloth, apply a little household cleaner, such as Formula 409, to the cloth and gently rub the surface. This will remove dirt and oily residues. Repeat on the back side of the hand. Dampen a cloth with water and wipe both sides of the hand, then thoroughly dry.

B. RESTORING HANDS:

1. Steel Hands:

Most clock hands are made of steel which has been heat treated to develop a deep blue or black surface color.

If there are significant areas of rust on the face of the hand, they should be removed in such a way that the hand is not scratched or otherwise marred. This can be done, using very fine wet-or-dry sandpaper (grade 600), or steel wool (grade 0000 or finer). Support the hand as in cleaning and rub carefully, lengthwise of the hand, until rust is removed.

This will leave areas of bright steel. Some collectors prefer to leave the hand this way. Others prefer a light coat of black spray enamel. The most authentic method is to apply heat to develop the original blue-black color in the metal. When doing this, be particularly careful to avoid getting fingerprints on the surface of the hands. Any oily residue will prevent coloring during heat treating.

a. Oven Method:

After the hands have been cleaned, place them on a piece of heavy aluminum foil on the broiler rack of your kitchen oven. Turn the heat to 550 degrees and allow the hands to "cook" for about a half hour. After about 20 minutes and every 5 minutes thereafter, look at them closely. When they have turned a dark purplish blue color, remove from the heat and drop in cold water. Thoroughly dry them.

b. Torch Method:

If you have a propane or butane torch, you can do this a bit more quickly. You will have to be careful, however, to stop the instant you obtain proper color.

The best procedure is to lay the cleaned hands on a piece of brass plate, or on a bed of dry sand, then apply the torch, moving back and forth along the hands to heat them evenly. When the desired color is reached, withdraw the flame immediately. Carefully handling them with a fork or tweezers, quench in cold water. Dry and polish with a soft cloth.

Another method, using a torch, is to hang the hand on a wire hook of sufficient length to avoid burning yourself. Slowly move the tip of the flame up and down the hand. Observe it carefully as the color changes. The instant the bare areas turn purplish blue-black, withdraw the torch and quickly dip the hand in water. Dry and polish with a soft cloth. This method results in more rapid color changes, so it must be done carefully to obtain proper color.

2. Brass Hands:

While brass hands do not rust, they do become dirty and can be cleaned in the same way as steel hands. In addition, because they are relatively soft, they may have suffered surface marking or scratching. If this is considered too unsightly, they can be polished, using very fine sand paper or steel wool as suggested for steel hands.

You may want to develop a higher luster, using liquid or paste brass polish, available in most supermarkets. Follow the manufacturer's instructions.

Chapter 2

CLOCK DIALS

In many cases, but far from all, the dial must be removed, after the hands have been taken off, in order to see the movement. For this reason, we discuss them here. There are essentially only two methods of attaching dials; to the case or to the movement. When the dial is attached to the movement, it is generally taken off after the movement is out of the case.

I. DIALS and RELATED COMPONENTS:

Since other elements are almost always associated with clock dials and sometimes are an integral part of them, let's look at some of them.

A. DIALS:

A dial is a usually flat surface, sometimes concave or convex, on which are placed circles, time divisions and numerals to facilitate the telling of time by the location of hands moving in front of it. Dials can vary in the materials used, the shape of both the edges and the surface contour, and the use of nonessential decoration. They usually have a hole in the center through which the hand arbor and cannon pass. Depending on the type of movement and the functions it performs, there may be additional holes for winding arbors, or for arbors controlling other actions.

1. Dials Can Tell More Than the Hour and Minute:

a. Seconds:

Many larger clocks, such as those designed to be hung on a wall, or to stand on the floor, have a third hand that makes one rotation a minute. This is referred to as a seconds hand. It usually rotates within a circle divided into 60 segments, but does not always advance at a rate divisible into 60.

Sometimes, a long hand is attached to an arbor concentric with the minute arbor and its outer end moves over the minute divisions of the time track. This is called a sweep second hand and requires no separate marking on the dial.

More commonly, a separate arbor on which a small hand is mounted, is located above the minute arbor and cannon tube. A small circle, with 60 divisions, over which the tip of the hand passes, indicates the seconds. This circle is called a "seconds bit".

An interesting variant is found in some Vienna Regulator wall clocks which have a conventional seconds bit (a circle with 60 divisions) and seconds hand, but the hand makes a revolution in much less than 60 seconds.

c. The Date, Month, Day of the Week:

An additional ring of numbers, from 1 to 31, on the dial make it possible for a third hand to indicate the day of the month. Sometimes, on the main dial or on an independent second dial, the day of the month, the name of the month and the day of the week are indicated. There are many variations.

d. Data Related to Time:

Occurrences that have a regular cycle, such as the phases of the moon, high and low tides and movement of the planets can be indicated by gearing added to the basic clock movement and shown in various ways, usually by adding additional indicating circles and hands to the dial that shows the time.

e. What the Clock Does:

You can usually tell what the clock does, by the number of winding holes in the dial plate.

(1) One Hole indicates that the clock only keeps time. It is a timepiece, not a true clock, since it has no strike train and does not strike the hours.

(2) Two Holes usually indicates that, in addition to keeping time, there is another train which strikes the hours. This, then, is a striking clock.

There is an exception to this rule. A few clocks use two springs to power a single time train, usually to provide longer running time, i.e., thirty days.

(3) Three holes are an indication that the clock has a third train, usually to operate chimes, playing melodies, such as the famous Westminster Chime.

2. Dial Plate Design and Construction:

The various materials used for making dial plates have influenced details of their design and construction. The availability of materials, over the years, in certain areas, has determined what clock makers used to make dials. We will discuss them in chronological order of their probable introduction.

a. Iron:

Wrought iron was probably the material used for dials on the earliest clocks. There was a resurgence of sheet iron dials in English country tall clocks during the 19th century and they were used, to a limited extent by American manufacturers in some of their iron-case clocks made late in that century.

Iron dials are comparatively thick and heavy and require painting or chemical treatment to protect them from rust. They are usually flat, although some American examples are cast with raised numerals and decorative features.

b. Wood:

Very early tower clocks and many later ones had dials made of wood. Because of its local availability, clock makers of the Black Forest in Germany used wood extensively for dials, as well as for movement parts, early in their history. In the early days of American mass production wood was the common material for dials and movements. There are still many such clocks available to collectors, so we will go into them in some detail later in this chapter.

c. Brass:

For many centuries, during the ascendency of handmade clocks, brass was the most commonly used material for both dials and movements. When polished it presents a most attractive appearance. It can be made soft enough to be easily engraved for decoration and it can be readily coated with silver or gold. Brass has been used extensively for dials on clocks made all over the world and is still common on dials for tall case clocks made today.

d. Zinc:

Zinc is a soft, easily formed metal that was used for dials on many millions of American clocks produced during the last half of the 19th and early part of the 20th centuries. Zinc dials are commonly found on Half-Column, Ogee, Column and Cornice and shelf clocks of various types. Because of its softness and the thin sheets used, many dials made of zinc have a concave ring inside the time track to provide desirable stiffness. Most zinc dials were painted, but some had a printed paper dial glued to the plate.

e. Copper - Enameled or Porcelained:

Many fine clocks have porcelain dials, notably those made in France, Germany and Austria (Vienna), as well as some of American make. Copper sheet is commonly used as the base material, to which is applied the liquid mineral porcelain mixture, which, when subjected to high heat, fuses to the copper to form a continuous glass-like coat. Numerals and other decoration are then added and fused to the background in the same manner.

f. Tin Plate:

This material, which is a thin sheet of steel with a very thin coating of pure tin applied to each side, has been used extensively in mass-produced clocks, particularly those made in America in the later years of clock making here. Tin plate of a given thickness is stiffer than zinc and less expensive. When used for clock dial plates, a printed paper dial was nearly always glued to the surface.

g. Aluminum:

In recent years, sheet aluminum has been used extensively by Korean clock makers for dials of their cheaper clocks, sold mostly through discount outlets here.

B. BEZELS:

A bezel is a decorative ring around the dial. It may be part of the dial plate, attached to the dial plate, or it may be a separate piece in front of the dial, fitted with glass and hinged to the case. Sometimes, there may be a bezel on the dial and another in front of it.

1. Bezel Attached to Dial Plate:

Frequently a formed brass ring is attached to a dial by means of ears passing through slots in the dial plate and bent over on the back side. Such rings are made of very thin sheet brass and are easily damaged, particularly when removed and not supported by the dial plate.

In better quality clocks, bezels attached to the dials are usually made of heavier cast brass, sometimes with decorative relief designs and sometimes machined and polished.

2. Bezel Separate From Dial Plate:

Many clocks have no bezel on the dial

plate, but have a hinged bezel into which glass is mounted. Such bezels hide the outside edge of the dial and, when closed, give the dial a finished look.

Separate bezels may be formed from fairly heavy to rather light brass sheet, to which the mounted glass gives substantial stiffness. Higher quality bezels are made of cast brass.

II. DIAL PROBLEMS AND SOLUTIONS:

A. ATTACHMENT to WOOD CASES:

1. Wood Dials:

PROBLEM: Many wood dials on American weight-driven clocks fit against the vertical wood strips that support the movement and are held in place by L-shaped nails driven into the face edge of the supporting strips. The dial rests on these nails and is held in place by turning the bent end over the face of the dial. If this action has been repeated often, the nail becomes loose in its hole and may have been lost.

SOLUTION: With a pair of pliers, pull the bent nail from the hole. Push the small end of a flat toothpick into the hole and break it off at the surface. Using the pliers, carefully push the nail back into the hole. Check to make sure that the bent end just clears the face of the dial as it is rotated.

If the original L-shaped nail is missing, it is a simple matter to make a replacement. Find a finishing nail that is a tight fit in the hole and press it in to the bottom of the hole. Put the dial in place and mark on the nail the location of the dial face, then remove the nail and make a 90 degree bend at that point. Cut off the bent portion so that it is about 1/4" long. Use pliers to push the pin firmly back into the hole.

2. Metal Dials:

Many metal dials on American weight clocks are held in place like the wood dials discussed above. The problems and solutions are the same as those for wood dials.

PROBLEM: Small, short screws used to hold the dial have little holding power and do not grip in the wood. Many dials of American clocks, particularly of the Kitchen variety, are formed like a pie pan, with the dial depressed below the rim, or with the dial soldered to the back edge of the bezel. They are attached to a thin wood piece which has been cut out to clear the depressed portion of the dial. The dial is then secured by small screws driven into the wood through holes in the bezel. Frequently, the cutout allows too little room for the screws, which break out at the inside edge of the wood. Because the wood is thin and the screws very small, they frequently become loose.

SOLUTION: If the hole is loose, put a tiny bit of glue on the end of a round toothpick and press it into the hole. After the glue has set, use a razor blade or sharp knife to cut it off flush with the surface of the wood. With an ice pick or other pointed instrument push a hole in the center of the toothpick and turn the screw into it.

In the case of a hole being broken out at the edge of the cutout, make a small block from a piece of wood the thickness of the dial support piece and about a half-inch long and shape it to fit the curve of the cutout. Glue in place so that the hole is about at the center of the block. After glue has fully cured, at least several hours and preferably overnight, put the dial in place and check to see if the holes in the bezel match those in the wood. If the bit of wood you have added prevents matching, carefully shave it with a sharp knife until the holes match. Install screws in the holes that match, then make a center hole in the piece you added and insert the screw.

B. CLEANING and/or RESTORING DIALS:

Before doing anything to a dial, careful consideration should be given to the effect such action may have on the relative value of the clock to a knowledgeable collector. In general, most collectors prefer that dials be preserved as they are, as long as they are not severely damaged. Careful cleaning is acceptable.

In this section, we will discuss dials according to their surface finish.

1. Exposed Metal Dials:

a. Brass, Plain, Silvered or Gilded:

Because they are subject to severe tarnishing, almost all such dials were originally finished with a coat of lacquer. It is usually sufficient to wipe them with a soft, damp cloth. If there are significant areas where the lacquer is missing and the metal has become badly discolored, the appearance may be improved by carefully using a silver polish, like that used on table wear, for silver plated dials. For brass dials, a brass polish, such as "Brasso", which is available in many supermarkets and hardware stores is a good bet. Follow directions carefully, and work only on the discolored areas. A very light coat of clear lacquer from a spray can will preserve the finish.

If there is extensive damage or deterioration of the dial, professional help should be sought.

2. Painted Dials:

a. Wood Dial Plate:

Most wood plates have a background of several coats of off-white paint on which details of the dial are painted. The numerals, Arabic or Roman, the time rings and their divisions nearly always consist of a single coat of thin black paint. Circles were usually scribed with a draftsman's ruling pen, as were straight lines of Roman numerals and minute lines in the time track. Commonly, these dials also carry colored designs at the corners and, sometimes, inside the time ring. gesso and gold leaf were frequently used to highlight details.

(1) Cleaning:

Because dial details consist of only a single coat of thin paint, occasional cleaning over the years has usually worn it away to one degree or another. Since the surface finish is dull, it tends to collect dirt and many old dials are very unsightly, so that cleaning is needed. In order to preserve as much of the original painting as possible, it is necessary to work carefully and slowly, a step at a time. After trying several methods, the author has found the following procedure most satisfactory.

Working on a well lighted table, prop the dial at a comfortable angle, by placing a support under the top edge. You will need a water color brush, cotton balls or swabs on a stick, a household detergent, such as Formula 409, or Fantastik, a soft cotton cloth (cotton Tee Shirt material is ideal) and a couple of fruit dishes, or other small containers.

Spray a little of the detergent solution into one of the containers, then wet the brush in it, wiping most of the solution out on the edge. Working with a nearly dry brush, choose an area about the size of a small postage stamp, near a corner of the dial plate and gently move the tip of the brush back and forth over it.

Place a little water in the second dish and moisten a small piece of cotton cloth, or a cotton swab, then wring it out and gently wipe the area just worked on. Turn frequently, to keep a clean surface in contact with the work. If it still appears dirty, you may want to go over it again. Remember that the objective is only to obtain a reasonably clean, uniform appearance, not a scrubbed look. Some discoloration is even desirable, as proof of age.

Continue this procedure over the rest of the dial, being particularly careful to use only the very tip of the brush around any black areas, which may dissolve and can smear the background. The author has found it best not to try to clean black surfaces, rather, very carefully use the tip of the brush to clean close to them, without touching the black.

(2) Restoring:

CAUTION: Before starting restoration work, read all of the instructions which follow. If you have any doubt about your ability to accomplish them, you may want to seek outside help.

If you have, and know how to use a draftsman's ruling pen on a beam compass, you may want to restore worn time ring circles. Tape a piece of stiff cardboard over the center hole in the dial and accurately establish the center of the rings.

With no ink in the pen, move it around the circle and adjust the center to correct for any deviation. Set your pen to match the width of the line you want to restore and then recheck to see that it follows that line closely. Shrinkage of the wood in the dial may have resulted in distortion of the circle, so that it cannot be accurately worked over. If the distortion is minimal, making a slightly wider line may be acceptable and may cover all of what remains of the old line. Use a grade of India ink made for use on mylar and proceed cautiously.

Straight lines can be restored, using a fine felt tip pen with permanent ink and a straightedge. Carefully outline Roman numerals, then fill them in. The curves of Arabic numerals must be carefully outlined free-hand, then the body filled in.

Colored and gold leaf decorations should be left as they are, unless you are skilled in such work. Fortunately, these elements are nearly always found in much better condition than those in black.

b. Painted Metal Dials:

(1) Cleaning:

Use the same methods described above for painted wood dials.

(2) Restoration:

If the background paint is in good condition, as it is with most iron dial plates, proceed as for painted wood dials.

Because of what engineers call a difference in the coefficient of expansion between the zinc plate and the paint film applied to it, changes in temperature have caused the paint on dials made of zinc to peel or flake off. The extent of this flaking can vary from just a few to many flakes. Unless a large area is affected, most collectors prefer to leave things as they are. Patching is possible by an expert, but usually the only solution is to strip all paint from the dial and completely refinish it. This is difficult and requires skills beyond the scope of this Primer.

People who do dial restoration and complete repainting regularly advertise in the MART of the National Association of Watch and Clock Collectors. For address, see the Introduction.

(3) Replacement:

Some clock parts supply houses and dial specialty houses stock new metal replacement dials. They usually cost much less than the charge for restoration of an old dial and you may want to substitute one for your old dial, temporarily, at least. By all means save the old dial and keep it with the clock.

c. Paper Dials:

A few paper dials were used on some of the earliest mass produced American wood movement clocks. They became common during the latter half of the 19th century and are found on a great many clocks prized by collectors today. Some have survived in very good condition, but more commonly, they show evidence of wear and tear.

Unfortunately, in many cases, clocks which originally had painted dials have had a replacement paper dial pasted over the original. Since replacement printed dials have been available for a good many years, some of these may give the appearance of age and the unwary collector may think they are original.

(1) Cleaning Paper Dials:

Paper is obviously soft and should never be subjected to liquids. The only satisfactory way of cleaning paper dials is to use a very soft eraser. We suggest either soft white ones, or art gum. Some plastic wall paper cleaners may work, too. Whatever you use, proceed carefully, a little at a time, starting at a place on the dial which is as inconspicuous as possible. Rub gently and watch for any sign that you are removing ink from printed areas, or damaging the surface of the paper.

(2) Restoring:

The only restoration possible with paper dials is to darken areas of numerals where ink has been worn away. This can best be accomplished with a black permanent ink felt tip marker having a fine point. If the surface finish of the paper has been worn away, there may be a tendency for the ink to run into the fibers, outside the desired area. Proceed cautiously, because there is no way of correcting errors, except by replacing the entire dial.

(3) Replacing Paper Dials:

(a) Ordering a Replacement:

Check suppliers' catalogs for a dial of the proper style and size to as closely as possible duplicate the original on your clock. In ordering, remember that the size of the dial is the outside diameter of the outermost circle, the time track. The new dial, as you will receive it, is printed on a rectangular sheet of paper. It has a cross printed to indicate the exact center of the time ring, but no holes for the hands or winding arbors.

The paper stock on which dials are printed varies in color and texture from shiny brilliant white to matte (dull) in colors ranging from pure white to various off-white shades, sometimes referred to as "antique".

(b) Grommets and Decorative Rings:

Most metal dial pans with applied paper dials have decorative thin brass rings, called "grommets" inserted into the holes for the winding arbors. They are secured by a number of sharp ears that project through the hole and are bent over in back of the dial plate. Using a knife, bend the ears up, one at a time, and remove the grommets. If the ears are broken, or the grommets missing, replacements can be obtained from clock parts suppliers.

Some dials also have a narrow brass decorative ring between the time track and the hand hole. These rings have several ears that project through slots in the dial plate and are bent over behind it. Gently bend them straight and remove the ring. Don't be surprised if some of these tabs break when you bend them. They are so thin that this is likely to happen.

(c) Preparing the Dial Plate:

If your old paper dial is complete, without holes or missing pieces and is securely in contact all over with the dial plate, the new dial may be applied over it. If there are holes, or pieces are missing, it will be necessary to remove the old dial from the plate. To do this, simply soak it in hot water, to which a little vinegar has been added to soften the glue. Peel all the paper off and scrub the surface with a stiff brush to make sure all unwanted material is washed off. Any unevenness will be reflected in the surface of the new dial.

(d) Cutting the New Paper Dial:

Dry the dial plate, then measure the diameter of the area into which the new dial must fit. It is a good idea to make a plain paper circle first, to check the fit. To do this, set a compass to exactly half the diameter measured and scribe a circle, then carefully cut on the line. See if the circle fits as you want it to. If not, adjust the compass and try again. When you have obtained the right size, place the point of the compass in the center of the cross on the new dial sheet and scribe a circle. Carefully cut out the new dial. Check again, to make sure it fits.

(e) Gluing the New Paper Dial:

The author prefers a yellow glue, such as Elmer's Carpenter's Glue for attaching dials, because it provides good bond and sets reasonably quickly. White glues may also be used. DO NOT USE CONTACT CEMENT OR RUBBER CEMENT.

Using a piece of stiff cardboard with at least one straight edge, spread a thin coat of glue over the back side of the paper dial, holding it firmly on a piece of clean paper so that no glue gets to the face. Make sure the entire surface is evenly covered.

Gently place the new dial face up on the dial plate and carefully position it, making sure that the numerals are properly placed with respect to the winding holes, with the figure 12 at the top. Holding it firmly in position with one hand, use the finger tips of the other hand to press it down, working from the center outward. Watch for bubbles and smooth them out toward the edge. Place a piece of waxed paper over the dial and, using a small piece of stiff cardboard with a straight edge, apply pressure to further smooth the paper dial. With a very lightly dampened cloth, wipe away any excess glue around the edge.

Place a piece of waxed paper over the new dial and weight it down with something that will accommodate the bezel at the edge of the dial pan and conform to the surface of the dial. A sack of sand is excellent for this purpose, or you can use a sack of sugar. Gently press this into place, being careful not to slide the dial out of position, then further weight it down with books or other heavy objects. It is a good idea to leave the weight on for at least a half hour, by which time the glue will have attained its initial set. Allow the glue to finish curing for at least several hours, preferably over night.

(f) Cutting Holes and Replacing Grommets:

When the glue has cured, it is time to cut out the holes for the hand shaft and the winding arbors. This is best done with a sharp-pointed hobbyists knife. With the dial side toward you, hold the dial plate so that light shows through the holes in the plate. With the point of the knife, pierce the center of the hole, then carefully cut around the inside. Use short strokes, cutting only on the down stroke. Repeat for other holes.

If your dial had an inside decorative ring, pierce the slots into which its tabs fit, then replace the ring, gently bending over the tabs at the back of the plate.

Replace the winding hole grommets, bending the retaining tabs over behind the plate.

d. Porcelain Dials:

In general, porcelain dials withstand ordinary wear and tear much better than any other type. They are usually found to be in excellent condition, even after more than a hundred years. Commonly used on European clocks, they are also found on many better American clocks.

Porcelain dials are, however, subject to damage from impact and sudden changes in temperature, which can cause cracking of the glazed surface. Fine cracks sometimes develop simply as a result of aging.

(1) Cleaning:

In most cases, simply wiping with a damp soft cloth will provide adequate cleaning. If not, you may use ordinary household window cleaning products. Proceed carefully, however, the numerals and other surface decoration is sometimes delicate and can be removed or damaged by cleaning materials. This is particularly true of maker's names, frequently added in paint to a stock dial.

(2) Restoring

Repair of damaged porcelain dials is difficult and demanding of special skills which are beyond the scope of a primer such as this.

Chapter 3

CLOCK PENDULUMS

In this chapter, we will discuss only the most commonly used type of pendulums, those which swing back and forth. A number of other types such as those that rotate, as in so-called "Anniversary" or 400-Day clocks have been used and are found on many clocks. The novice, however, would be well advised to practice repair of the swinging pendulum type before proceeding to others.

Pendulums are frequently, but not always, located in front of the movement. When this is the case, they are readily visible after the dial has been removed. When the pendulum is behind the movement, it may be supported on the back plate of the movement, or on the back of the case.

Since it was discovered that a pendulum of a given length always swings at the same rate, that is, makes the same number of swings in a given time period, pendulums have been used to control the rate at which a clock runs. If a pendulum were swinging in a vacuum, it would continue to oscillate forever. Air, however, offers enough frictional resistance to eventually cause the pendulum to stop swinging.

For a pendulum to be effective in controlling the rate at which a clock runs, some means of adding power to keep it moving must be provided. The weights or springs that power the movement provide this power by giving the pendulum rod a very slight push at the end of each swing.

An important point to remember is that the rate at which a pendulum oscillates (swings) is dependent on its length. The longer the pendulum, the slower it will swing. The shorter the pendulum, the faster it will swing.

To understand how a pendulum performs in a clock, let's take a quick look at a few details and terms commonly used by clock makers.

I. PENDULUM ELEMENTS:

As used in clocks, the pendulum consists of four basic elements, a flexible suspension device, a rod, a bob and a means of finely adjusting its effective length.

A. Pendulum Rod:

The rod is a stiff piece of material that is supported at its upper end and carries a weight, usually a bob, near its bottom end. It may be made of wood, steel, brass, or a combination of such materials.

B. Pendulum Bob:

The bob is nearly always made of metal, but may take a great many shapes. Its function is to provide a concentrated weight, the center of which (the center of gravity) determines the effective length of the pendulum. Moving the bob up shortens the effective length and causes the pendulum to swing faster. Lowering the bob lengthens the effective length and makes it swing at a slower rate.

C. Length Adjustment:

There are two basic methods of

adjusting the effective length of the pendulum. One raises or lowers the bob on the rod, by means of a threaded element which is part of, or attached to the lower end of the pendulum rod and a nut on which the bob rests. The other consists of a fork fitting over the suspension spring which is raised or lowered by a screw to change the effective length of the pendulum.

II. PENDULUM RODS:

A. Steel Wire Rods:

Probably the most commonly found type of pendulum rod on American clocks, steel wire rods are also found on many European clocks. Generally, they are made of steel wire around 3/32" in diameter, #14 gage, or larger. While employing wire for most of the length of the pendulum, they differ in the means of attachment at both top and bottom.

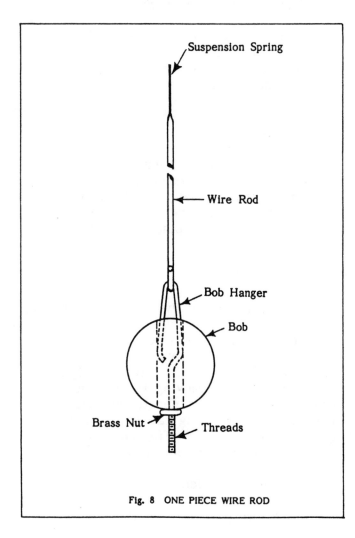

Fig. 8 ONE PIECE WIRE ROD

1. One Piece Wire Rods:

The simplest type of pendulum rods are made from a single piece of steel wire and include both the suspension spring and a hook for supporting the bob. They were used in great numbers by American clock makers from the beginning of mass production and continued to be used for almost a hundred years.

a. Construction:

After being cut to the proper length, about 1 3/8" of one end of the wire was flattened to a thickness of about .005", (five one-thousandths of an inch), by squeezing and drawing it between steel rollers. This flattened area became the suspension spring for the pendulum. A small dimple punched near the end served to hold it in place when placed in the slot of the supporting block on the movement. At the opposite end of the rod, a small hook, bent in the same plane as the suspension spring, was formed to support the bob.

b. Problems:

The comparatively delicate suspension spring portion of these pendulum rods is easily bent by improper handling. When this happens, the pendulum will wobble and not properly perform its function.

The rod itself is made of soft steel and may be easily bent, affecting performance. It should be reasonably straight. If it is bent, it can usually be straightened with the fingers.

c. Correcting Problems:

(1) Repair:

Unhook and set the bob aside.

Carefully remove the pendulum rod from the slit in the block on the movement that supports it by grasping it with the fingers, underneath the support block, while placing a fingernail behind the part projecting above the block, then gently pull forward.

Bent rods can easily be straightened

by applying pressure with the thumb, while supporting the rod between two fingers. Do this a little at a time, sighting along the rod and repeating until it is straight.

If the suspension spring portion of the rod is bent, it may be possible to make it usable. If it has been bent so that a sharp crease mark is visible, gently grip the spring with a pair of flat face long nose pliers placed so that one edge of the jaws is right alongside the crease mark. With the tip of a finger held as close as possible to the plier jaws, gently press the spring to correct the bend. Check visually to see if the spring is flat and repeat, if necessary.

If there are no visible creases, but the spring is twisted or curved, hold it between the thumb and two fingers of one hand and, applying slight pressure with the thumb, use the other hand to pull it between the thumb and fingers, in the direction of the curve. Repeat until it is as flat as you can get it.

Carefully replace the spring in the slot of the support block, replace the bob and set in motion. If it swings without wobbling, you have made a successful repair. If not, remove it and repeat the procedures. If you find it impossible to obtain proper motion, it will be necessary to replace the rod.

(2) Replacement:

Almost identical one piece rods are usually available from clock parts supply houses.

Measure the new rod against the old one, bend a hook at the same point as on the original, then cut off the excess rod. Make a dimple at the top of the spring by placing it on a piece of wood or hardboard and, using a center punch, or the tip of a small nail that has been rounded slightly by filing, gently tap with a small hammer. Be careful to make sure that the dimple is centered in the width of the spring.

2. Multiple Piece Rods:

Some pendulum rods are made up of two or more pieces, permanently or loosely attached to each other.

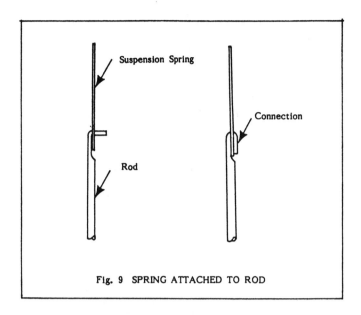

Fig. 9 SPRING ATTACHED TO ROD

a. Attached Suspension Spring:

The simplest form of multiple piece rod, has a separate suspension spring permanently attached to its top, with a hook for the bob at the bottom. A short portion of the top end of the rod is flattened, run through a hole in a flat spring and bent over to secure it. In effect, it is like the one piece rod, subject to the same kind of damage and repair, except that the spring is hardened and less easily damaged, but also less easily straightened if it has been bent.

b. Two Part Rod: Figure 10

To facilitate easy removal of the bob, many pendulums are made in two or more pieces, with hook connections between them as shown in Fig. 10. Some connections, particularly on recently made tall clock movements, are relatively intricate. Methods of disconnection are usually illustrated in the Owner's Manual supplied with the clock.

c. Composite Rod: Figure 11

In this form, the rod has threads at each end. A threaded brass block with hook for attachment to the suspension

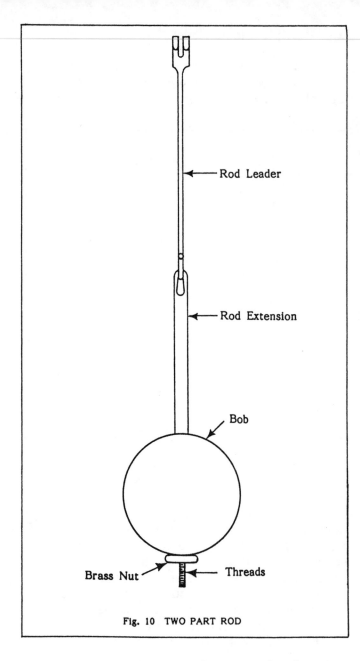

Fig. 10 TWO PART ROD

Rod Leader

Rod Extension

Bob

Brass Nut — Threads

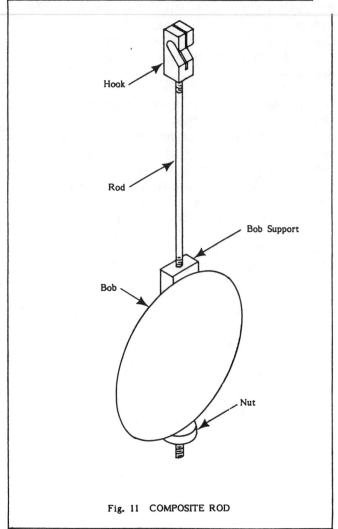

Hook

Rod

Bob Support

Bob

Nut

Fig. 11 COMPOSITE ROD

spring is screwed on at the top. At the bottom end there may be threads onto which a brass bob support is screwed, or there may be a loop formed to accept a flat hook at the top of a separate bob support.

3. Gridiron Pendulums:

True Gridiron pendulums consist of several alternating brass and steel rods attached in a manner designed to compensate for differences in length caused by changes in temperature. Many pendulums have been made to look like Gridirons but do not perform their compensating function.

B. Wood Pendulum Rods: Figure 12

Wood rods are commonly found on larger wall clocks and some tall clocks. They are nearly always painted or varnished to help reduce the penetration of moisture, which causes wood to shrink or swell.

1. Construction:

Wood rods are usually of oval or rectangular cross section and may have a metal hook attached to the top. This may be of formed sheet brass or steel, which wraps around the top of the rod and is attached to the rod by crimping. Alternatively, it may be a brass piece which has a flat at the bottom, fitting into a slot or mortise in the rod and held in place by screws or rivets.

A short threaded steel rod is attached to the bottom of the wood rod onto which

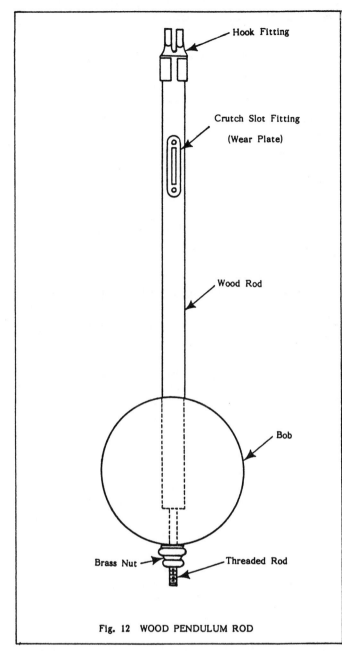

Fig. 12 WOOD PENDULUM ROD

b. Short Wood Pendulum Rod Extensions:

Some smaller wall clocks and some table clocks may be found with short oval shaped wood pendulum rods, acting as extensions of wire rods working in conventional slotted crutches. They differ from those discussed above in that they do not have a slotted wear plate to accept a crutch pin. The wood extension rod is suspended on a hook or T-bar at the bottom of the wire rod.

2. Problems with Wood Pendulum Rods:

The most common problem with these rods is that the attached hardware can become loose. We will discuss the various fittings, one at a time.

3. Correcting Problems:

a. Top Hook Fittings (Hangers):

(1) Formed Sheet Metal: (Fig. 12)

Most commonly, these are made of relatively thin sheet metal, formed to wrap around the rod. The ends have one or two small ears, bent so that they dig into the wood. When they have become loose, use pliers to gently squeeze the fitting at each edge where it is bent to wrap around the rod. The idea here is to bend the metal near where the original bend was made, at the edge of the rod, so that pressure is exerted over the entire surface.

Check the hook to make sure that it fits properly over the suspension spring pin, or, in the case of a short pendulum, in the ring at the bottom of the wire pendulum rod. If the hook is split, that is, has a slot into which the suspension spring fits, make sure that both parts of the hook have identical shape and properly fit the support pin well. If they do not, use long nose pliers to carefully reshape them.

(2) Flat Metal Plate:

Usually thicker than the sheet metal variety, these fittings may fit into a slot cut in the wood rod, or onto a flat face of the rod. They are attached by

is threaded an adjusting nut to support the bob and provide a means of moving it up or down.

a. Wood Pendulums in Wall and Tall Clocks:

These pendulum rods are nearly always oval in shape and painted black. At a point a few inches below the top a short, narrow slot is cut into the center of the rod. A brass plate with a somewhat narrower slot is located over the slot in the rod and attached by screws or rivets. A metal pin on the crutch of the movement fits into this slot to impart power to the pendulum.

means of screws or rivets, which may have become loose. Try carefully tightening screws with a small screwdriver. If the screws are loose and will not grip, remove them and insert the end of a round toothpick with a dab of glue on it. Allow the glue to set, then cut the toothpick at the surface of the rod, using a razor or sharp knife. Make a center hole with the tip of a sharp instrument and carefully drive the screw into place.

If rivets have been used for attachment, and have become loose, place the rod so that the head of a rivet is in contact with a hard surface, then tap the other end of the rivet lightly with a small hammer, until it grips tightly.

b. Crutch Slot Fittings (Wear Plates):

The purpose of these fittings, nearly always made of brass, is to provide a slot of accurate width into which a pin on the end of the crutch is inserted. The clearance between the pin and the edges of the slot is important for proper functioning of the clock. A steel pin in a brass slot minimizes friction and resulting wear. For this reason these fittings are referred to as "wear plates".

Wear plates are nearly always mounted on the front face of the pendulum rod. They must be firmly attached in order to effectively transmit power from the crutch to the rod.

If rivets or screws have become loose, follow the procedures suggested for top hook fittings, above.

(1) Replacement:

If the slotted plate is missing from your pendulum, a new one may be obtained from some clock parts supply houses.

c. Pendulum Adjusting Fittings:

These are very finely threaded metal rods, with a nut, attached to the bottom of the wood pendulum rod. The means of attachment may loosen, or the nut may be stripped. In our discussion here, we will refer to the threaded rod simply as a screw.

(1) Screw Threaded into Rod:

Particularly in clocks of recent manufacture, the screw is simply threaded into a small hole in the end of the wood rod. If it becomes loose, to the point where it will not support the bob, place a dab of glue on its end and push it into the hole. Allow several hours for the glue to set before hanging the bob on it.

(2) Riveted Attachment:

There are a number of ways in which adjusting screws are attached to wood rods, using rivets, which frequently become loose. Since there is little if any movement encountered by the rivets during operation of the clock, this may not be a condition requiring repair, if the bob is adequately supported. It is, however, a simple matter to tighten the rivets by supporting the heads on a hard, solid surface and lightly tapping the opposite end with a small hammer.

B. Solid Brass Pendulum Rods:

Commonly found on English and some other movements, these rods are made of rectangular brass, usually about 1/8" x 5/16" and have no hooks. At the top end is a slot with a small hole drilled transversely through its sides. The suspension spring fits into this slot and is held in place by a taper pin inserted through it and the slot holes.

A short way down from the top of the rod, a narrow slot is cut out to receive the crutch pin.

At the bottom end, the rod is turned to form a thin, round section, which is threaded to accept an adjusting nut. This nut is usually provided with a tapered section at the top which fits into a matching slot in the bottom of the bob. This insures both upward and downward movement of the bob, as the nut is turned.

C. Pendulum Rod Leaders:

Commonly used on movements of comparatively recent German manufacture, leaders may be found in conjunction with many types of pendulum rods. They provide

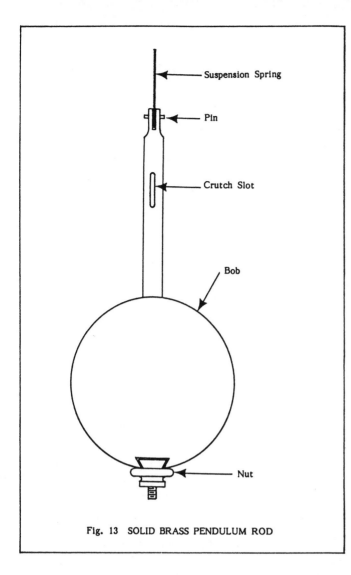

Fig. 13 SOLID BRASS PENDULUM ROD

Suspension Spring

Pin

Crutch Slot

Bob

Nut

a link between the suspension spring and the rod, or may act as a rod, themselves. Usually made of stamped sheet brass, the top attaches to the suspension spring, while the bottom is formed in many different ways to attach to a pendulum rod, or directly to a bob. There may be an intricately formed section about midway of the leader into which the crutch fits, sometimes including a mechanism for adjusting the beat. (See Chapter 5)

In its simplest form, the center section of the leader is straight and a slot in the crutch acts against its sides to impart power to the pendulum.

1. Problems with Leaders:

While they may become bent due to severe mishandling, there is little else which can go wrong with leaders.

a. Disconnecting the Pendulum Rod:

Disconnecting the pendulum rod, can pose a problem, if you are doing it for the first time. If you have access to an Owner's Manual for the movement you are working with, you will usually find specific instructions. Some of these leaders are designed to lock the pendulum rod onto the leader, when it is in place. With this type, it is necessary to slightly lift the rod, while tilting the bottom toward the back of the clock, in order to completely unhook it.

III. CLOCK PENDULUM BOBS:

For this discussion, we will refer to any device that acts as a weight to define the center of gravity of the pendulum assembly as a "Bob". Bobs may vary in design from a simple lead or iron disc with a steel loop embedded in it as an attachment, to a hook in the rod, to complex assemblies.

While sometimes referred to as "pendulum balls", we prefer the more generic term "bob", since often they bear no resemblance to a ball.

To permit adjustment of the effective length of the pendulum, necessary to achieve accurate time keeping, a means of varying this length must be provided. Sometimes this is done by changing the effective length of the suspension spring, but more frequently it is accomplished by providing a mechanism for raising or lowering the bob on the pendulum rod.

In its simplest form, definition of the center of gravity of the pendulum is the bob's primary function. Most bobs, however, are designed to at least partially provide compensation for changes in the length of the pendulum assembly resulting from changes in temperature. For example, an increase in temperature which causes the rod to lengthen, will also cause the bob to expand. When it is supported at its bottom, such expansion is upward and to some degree compensates for the downward movement of the pendulum rod. The larger the bob, the greater the movement, when expansion occurs, more nearly achieving the goal of maintaining a constant center of gravity.

Some bobs are very light in weight, while some are very heavy. There is also wide variation in size. The reasons are obscure and we won't go into them at this time.

A. Types of Pendulum Bobs:

1. Non-Adjustable Pendulum Bobs:

Usually made of cast iron or lead, these bobs are commonly found on American Tambour case (Humpback or Camelback) clocks and in some others. They have a wire loop for attachment to a hook on the bottom of the pendulum rod. They may be of any shape, often have elaborate raised designs and sometimes the initial or trademark of the maker. They may be painted, usually black or gold, or may be plated.

These bobs have no provision for raising or lowering them on the rod. Adjustment is accomplished by means of a device attached to the back plate of the movement, which can be activated by a key from the front of the movement.

a. Remote Adjustment Mechanism:

This mechanism includes a U-shaped bracket attached to the back plate of the movement, with a slot to support the suspension spring. An L-shaped fitting with a slot for the suspension spring and another into which fits a shouldered screw, secures it while allowing movement up or down. A small threaded rod with pinion gear at its top goes through a hole in the top of the support bracket, then is threaded through the slot fitting and its lower end is retained in a hole in the bottom of the support bracket. A small arbor with a pinion engaging at right angles to and mating with the pinion on the threaded rod and a square section for a key at its other end, goes through the back and front plates of the movement.

Turning the arbor with a key turns the threaded shaft, thus raising or lowering

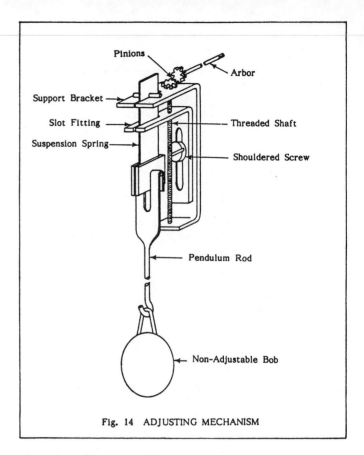

Fig. 14 ADJUSTING MECHANISM

the slot fitting. The working length of the suspension spring, and of the pendulum, is lengthened or shortened and the rate at which it swings slowed or speeded up. A hole in the dial, usually above the figure 12, allows the key to be inserted. On either side of this hole a letter S and F indicate the direction in which the key must be turned to make the clock run slower (S) or faster (F).

2. Adjustable Wire Loop Pendulum Bobs:
    (Fig. 8)

Adjustable wire loop bobs are commonly found on American weight-driven clocks, early shelf clocks and others.

This type is usually round, convex on the front and flat with a rectangular block on the back. They are made of lead with a thin brass sheet forming the front face and crimped over the back edge. The block on the back has a rectangular hole through which a bent wire loop with a threaded tail is inserted. The loop provides a means of attachment to the hook on the bottom of the pendulum rod. A small, round brass nut is threaded on the projecting tail, to support the bob and provide a means of raising and lowering it to adjust the rate of beat.

Size ranges from 1 1/2" to 2" in diameter.

a. Problems with Adjustable Wire Loop Bobs:

The most common problem is that threads in the nut become stripped, so that it will not stay in place. while nuts alone are not generally available, a replacement assembly, with a nut on it is available from parts supply houses, at small cost.

The wire is bent so that the sides of the loop exert slight pressure on the sides of the hole in the bob. If they do not, the bob may wobble. This is easily corrected by removing the nut, taking the wire out of the slot, from the top, bending the loop slightly outward and re-attaching the nut.

Because of this slight pressure on the sides of the hole, the weight of the bob may not be enough to overcome friction at that point, so that it won't move down by itself when the nut is lowered. When adjusting the bob to slow the clock, it is a good idea to pull the bob gently downward to make sure it is in contact with the nut at its new location.

3. Sliding Pendulum Bobs:

Whether the pendulum rod is of wood, brass or steel, the most common bob is of this type, designed with a hole the shape and size of the rod cross section, or, in some cases, that of a rod extension, so that it can move up or down on the rod. The bottom is supported by a nut threaded onto the rod end, or a rod extension.

Sliding bobs are used on a wide variety of clocks and vary considerably in size and shape. They may be made entirely of sheet brass, convex, or lens-shaped on both sides; with the front side of brass and the back of steel or plastic; of lead with brass facing; of cast iron or brass with complex designs; of various materials, as with mercury types and simulations thereof; even of wood, as with

cuckoo and other German Black Forest clocks.

a. Two-Piece Bobs for Wood Rods:

Extensively used on modern German movements, particularly in tall-case clocks, these pendulum bobs are of comparatively light weight, with brass fronts and steel or plastic backs. The brass is crimped over the steel or plastic back which has a cutout at the top of sufficient size to allow the rod to pass through it. At this opening, the back lip of the brass front is bent to the shape of the rod, usually oval. An ear is formed in the steel back and bent in toward the front in such a way as to hold the rod in place against the formed section of the brass front, thus keeping it from wobbling. This ear should just barely touch the wood rod, so as to allow the bob to slide freely, yet maintain its position parallel to the line of the swing.

At the bottom of the bob, another, smaller, opening is provided, with a depression in the lip of the brass rim, intended to hold the threaded extension rod in position. A round nut on this rod retains the bob and provides for adjustment. When adjusting the bob downward to slow the clock, it is a good idea to press down gently on the pendulum bob to make sure its bottom is in contact with the nut in the new position.

b. Solid Bobs for Oval or Rectangular Rods:

Many pendulum rods are made of oval or rectangular brass or steel, which are usually of one piece, but may be an extension of other types of rods. Bobs for these rods are usually solid, made of cast brass or plated pot metal in many shapes and sizes, or, in the common lens shape, of lead with brass facing. All have a hole of the shape and slightly larger than that of the rod which passes through them.

Oval rods are commonly threaded to receive the adjusting nut. Because of their shape, the threads appear only on the outer edges of the rod.

Rectangular rods usually are turned down to a round cross-section at the end, before threading, or a steel threaded rod is inserted. They have full threads.

(1) Problems and Solutions:

Not infrequently, the threads on the edge of an oval rod are very shallow, have very small surface area and cause the nut to fit very loosely, sometimes to the point of not moving when turned, or capable of being removed, without turning.

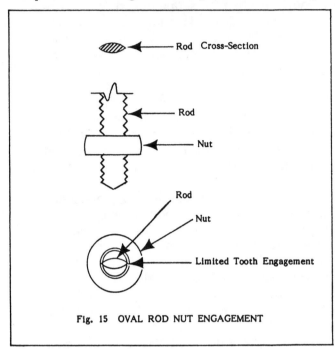

Fig. 15   OVAL ROD NUT ENGAGEMENT

Sometimes, this situation can be corrected by slightly widening the rod in the area where the nut operates. With the bob and nut off, lay the threaded end of the rod on a solid steel surface, with the threads to the sides and gently tap the face of the rod several times with a light hammer. This will expand the metal outward. Try the nut and repeat the procedure until a snug engagement is obtained. If this fails, it will be necessary to obtain skilled professional help, or to replace the rod and nut.

c.   Cuckoo Clock Bobs:

Cuckoo clock bobs are unique in that they are generally made of wood, in a variety of shapes from round to simple or elaborate carved leaves or bunches of grapes.

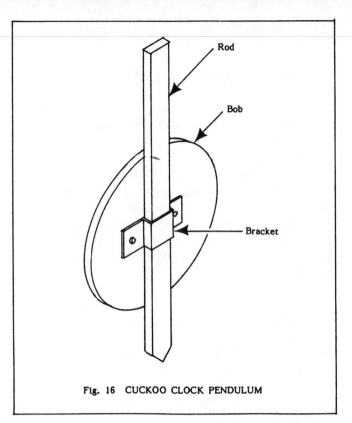

Fig. 16   CUCKOO CLOCK PENDULUM

The pendulum rods of larger cuckoo clocks are nearly always made of rectangular shaped wood. Nailed to the back of the bob is a thin strap of steel, bent to fit the rod in such a way as to exert pressure on it, but to permit movement of the bob on the rod. Friction holds the bob in place, after adjustment.

(1) Problems and Solutions:

The metal strap may become so loose as to not hold the bob in position on the rod. If this happens, remove the bob and, using a screwdriver, press gently against the center of the strap to bend it toward the back of the bob. Try it on the rod and, by bending, adjust it so that it stays in place, but can be rather easily moved.

d.   French Clock Bobs:

The pendulums of French clock movements are nearly always suspended from the back plate of the movement.

Occasionally, as in the case of Empire column clocks, the pendulum is easily accessible from the rear of the clock. More commonly, there is an opening in the back of the case of the same diameter as the front bezel opening and access to the pendulum is made difficult. This situation is further complicated by the fact that the bell for striking the hours is mounted in back of the pendulum and must usually be removed before the pendulum can be disconnected. The bob is only accessible after the pendulum is removed from the case.

Many French clocks employ bobs with adjustment nuts located within, rather than at the bottom of the bob. Round bobs have a horizontal slot in the middle to accommodate the nut. There is also a small set screw that secures the bob to the rod, after it has been adjusted. Before adjustment, this screw must be loosened and then retightened after adjustment of the bob. With complex composite bobs, the nut may be located slightly above what appears to be the bob.

Frequently, an adjusting mechanism is provided at the suspension spring, which is usually adequate. Change of the bob position on the rod is seldom needed.

When necessary, adjustment of the bob is accomplished in the same way as it is when the adjusting nut is located under the bob, by turning the nut. Remember to loosen the set screw, before adjustment, and to retighten it after adjustment is complete.

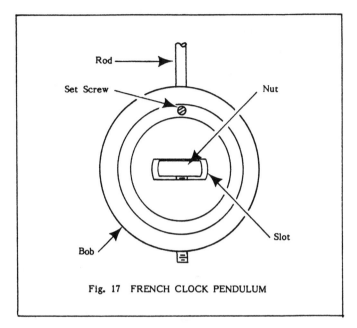

Fig. 17 FRENCH CLOCK PENDULUM

Chapter 4

CLOCK PENDULUM SUSPENSION DEVICES

The purpose of a pendulum suspension device is to support the pendulum, while allowing it to swing with a minimum of resistance and to do so for a long period of time, without failure. In addition, it must keep the pendulum swinging in a straight line, without wobble. Some such devices also are equipped to alter the effective length of the pendulum, thereby providing a means of adjusting the rate of pendulum oscillation. (Fig. 14)

Pendulums may be suspended from the front or back plates of the movement, or from the back board of the case. A pin or block of brass or steel is used to provide this support. Design details will vary with the type of suspension.

I.   SPRING SUSPENSION:

A.   SINGLE SPRING SUSPENSION:

Single suspension springs are made of relatively thick material and, to provide adequate flexibility, are comparatively long. They are commonly used with long pendulums, particularly as in tall-case clocks, and in many types of American, German and Korean wall, shelf and mantel clocks.

1. Integral Single Suspension Spring:
   (Fig. 18a)

As we have pointed out, the suspension spring is an integral part of some pendulum rods. It is simply a flattened end of the same piece of steel from which the rod is made. In effect, it is a single flat spring of sufficient width and thickness to give torsional stability, that is, to keep the pendulum swinging in a flat plane. It is inserted in a slot in

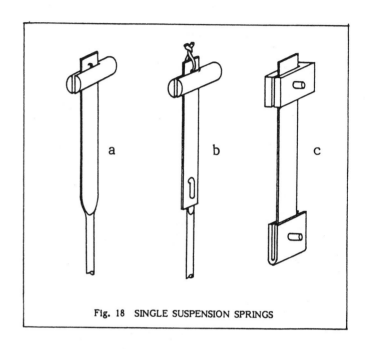

Fig. 18   SINGLE SUSPENSION SPRINGS

a support device and held in place by a dimple near its end, which will not pass through the slot.

a.   Support:

With both wood and brass movements, the support for integral suspension springs nearly always consists of a brass pin set into the front plate of the movement and secured by riveting. A very narrow vertical slot is sawed into the pin to receive the suspension spring.

2. Permanently Attached Single Suspension Spring: (Fig. 18b)

In this form, a separate piece of flat spring steel is permanently attached to the rod, the end of which is flattened, then passed through a hole in the spring and crimped over. It is retained by a twisted wire through a hole at the top.

a. Support:

In nearly all cases, the support for this type of suspension consists of a brass pin with a vertical slot, mounted either on the front or back plate of the movement.

3. Separate Single Suspension Spring:
(Fig. 18c)

Commonly, these springs are made of a strip of flat spring steel, from about 3/16" to 3/8" wide and from 1" to 3" long, with a hole near one end and a folded brass clip at the other, through which a short pin is fixed. The hole end is inserted in a narrow slot in the support bracket and held in place by a taper pin run through holes in the bracket and the hole in the spring. The pendulum hook fits over the pin on the bottom end.

a. Support:

Support for this type of spring may be on the front or back plate of the movement, or on the case back. It is usually a slotted block of brass, but may be a slotted pin.

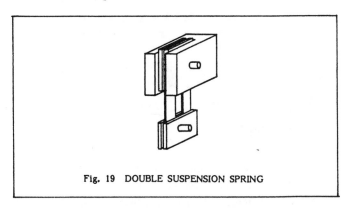

Fig. 19 DOUBLE SUSPENSION SPRING

B. DOUBLE SUSPENSION SPRINGS:

These springs, consisting of two narrow, thin and short springs mounted in brass or fiber are found, in a variety of forms on a great many clocks.

Double suspension springs can be much shorter than typical single suspension springs, because the spring material is much thinner and, thus, more flexible. Stability is achieved by providing a space between the two strips of spring, thus giving the effect of a much wider spring.

Older original double suspension springs had two small brass plates at the top and bottom of the assembly, between which the two spring strips were sandwiched and riveted to secure them. A small hole is provided in the top sandwich to receive a taper pin run through it and holes in the support block. A short pin is attached at the other end to receive the pendulum hook.

In clocks of recent manufacture, the top and bottom plates of these springs are usually made of fiber, rather than brass. This is also true of most replacement springs currently available.

1. Problems and Solutions:

These springs are comparatively delicate and are likely to break if they are severely bent. Many times, only one of the two spring strips will break. The remaining one will probably support the pendulum, but it will wobble as it swings. Replacement will be necessary.

2. Replacement:

Replacement suspension springs are available from several clock parts suppliers, whose catalogs illustrate in actual size the many types and styles you may encounter.

Because there are a great many sizes and shapes of these suspension springs, it is important to order a new spring that matches the old as closely as possible. To do this, place the old spring over illustrations in the catalog until you locate the proper one.

If your old spring has brass plates and the new one has fiber plates, it is likely the new one will not be as thick as the old one. Check this by carefully inserting it in the support slot. If it is very loose, use a pair of pliers to squeeze the sides, narrowing the slot so that it is only a little wider than the suspension spring plate.

II. THREAD SUSPENSION:

Some clocks, particularly those of early French manufacture, suspended the pendulum on a loop of silk thread. While they are not very common in this country, there are a number of them in the hands of collectors, and you may come across one occasionally.

In this system, a short length of silk thread is supported at each end by being threaded through holes in a support pin and knotted to retain it. This forms a loop on which the top hook of the pendulum rod is hung. It operates very well, but the silk does deteriorate and must be replaced occasionally.

A. Repair - Replacement:

Nylon thread is a good substitute for silk. Simply make a loop of suitable size and secure the ends above the support pin.

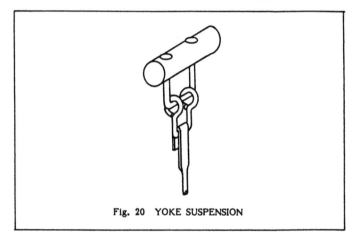

Fig. 20   YOKE SUSPENSION

III. YOKE SUSPENSION:

Most cuckoo clocks employ a yoke suspension system. It consists of a U-shaped wire yoke hung parallel to a supporting rod, with another U-shaped wire yoke, having eyes at the ends, hooked over it, to make what looks like a tiny swing. A hook on the end of the pendulum rod fits over the second yoke. When the pendulum is in motion, the rings of the lower yoke simply rock on the bottom of the supporting yoke.

A. Problems and Solutions:

1. Rust:

Because they are made of steel wire, the parts of this type of suspension can become rusty, with resultant friction that affects their performance. If this occurs, try using a thin strip of crocus cloth, or very fine. 600 grit wet-or-dry sand paper, both of which are available at stores carrying auto paint and refinishing materials. Rub lightly until all trace of rust is removed and the surfaces are polished.

2. Bent Yokes:

The upper yoke should be perpendicular to the supporting rod. If it is not, use a pair of pliers to carefully straighten it.

The rings of the lower yoke should rest on the flat part of the upper yoke, just inside of the point where they begin to curve upward on each side. With pliers, or a screwdriver, carefully squeeze or spread the sides, until this condition is achieved.

IV. KNIFE-EDGE SUSPENSION:

Rarely encountered by the average collector, this type of suspension was used on early clocks with crown-wheel (Verge) escapements. It has also been used on some very high quality astronomical and other specialized clocks.

On the end of the pendulum is mounted a sharp V-shaped piece of hardened steel or gemstone which rests in slightly more open V-shaped grooves in the supporting block. This results in a minimum of friction and insures pendulum stability. A disadvantage, however, is that dust accumulates in the groove and can cause wear, unless the movement is mounted in a hermetically-sealed case, as is usual with very accurate clocks of the type that would use this suspension.

Chapter 5

EXAMINING MOVEMENTS IN THE CASE

and

CORRECTING SUPERFICIAL PROBLEMS

Frequently, problems capable of causing the clock to stop can be detected and corrected while the movement is still in place in the case. This may be true, even with movements that have the dial attached to them, when the back has a large door giving access to the movement, or where the backboard can be removed. So we will start by looking for evidence of malfunction that can be seen with the movement in place in its case.

I. BACK or SEAT BOARD MOUNTED MOVEMENT:

Movements which are attached to and supported by the back of the case, or which are attached to a seat board supported at the sides, nearly always have dials attached to the case, not the movement. The pendulum is suspended from the front plate of the movement.

In Chapters 1 and 2, we found how to remove the hands and dial. When this has been accomplished, the front of the movement, the suspension, the pendulum and the crutch can be seen, as well as other features. Let's check some of the things we can now see, that might be causing problems.

A. The ESCAPE WHEEL:

Carefully examine the escape wheel teeth for any irregularities. If the tips of the teeth are bent over, even slightly, that poses a problem which can only be dealt with after the movement has been removed from the case. See Chapter 8 for repair procedures.

CAUTION: DO NOT WIND THE CLOCK if escape wheel teeth are bent. Further damage may result.

B. The SUSPENSION SPRING:

If the suspension spring is bent or broken, repair or replace it. (See Chapter 4.)

Is it properly located in the support? If it is retained by a dimple near its top, the dimple should be in contact with the top of the support and the back edge of the spring should be at the back of the slot. The front of the slot should be tight enough to keep the spring in place, but the section of the slot in which the spring is suspended should have enough clearance so that the spring can move to permit the pendulum to hang plumb when the bob is in place.

Occasionally, the dimple at the top of the spring is not deep enough to prevent the spring from sliding through the slot. If this is the case, carefully remove the bob, then the rod. Being careful not to damage the spring, lay it so the spring is flat on a piece of soft wood with the hollow side of the dimple up. Round the point of a small nail, using a file or sandpaper, then place the point in the dimple and tap lightly with a small hammer. This should increase the depth of the dimple sufficiently to retain it in the slot.

Sometimes, you may find a hole, rather than a dimple in the suspension spring (Fig. 18b). If so, there should be a fine wire run through the hole and twisted at the top to retain the spring. If this wire is missing it can be replaced with #26 Black Annealed Iron Wire, available at most hardware stores. Cut a piece about two inches long and run it through the hole in the suspension spring. Bend sharply at the center then twist the ends two or three times to secure it. Snip off the excess. If you don't have wire cutters, scissors will do, but we don't recommend using the ones you or your spouse use in sewing.

C. The PENDULUM:

1. Wire Pendulum Rod:

Is the rod reasonably straight? If not, it should be removed and straightened as suggested in Chapter 3.

Does the bob hang so that it is parallel to the face of the movement? If not, take the bob off of the rod, remove the rod from the movement and hold it tightly in a vise, near the spring end, then grasp the hook with a pair of pliers and twist slightly until it is parallel with the suspension spring.

Replace the rod in the suspension support, making sure it goes through the loop in the crutch, attach the bob and set the pendulum in motion. If the clock ticks, you have solved a problem. Adjustment of the bob by raising or lowering the rating nut will be probably be necessary to obtain good time keeping.

2. Other Pendulum Rods:

Is the rod properly attached to the suspension spring? If it is not, reattach it, making sure the crutch is properly engaged, then set the pendulum in motion. This may have been the reason the clock stopped.

If yours is a wood rod, check to see that the top hook fitting and the crutch slot fitting are securely attached. A loose crutch slot fitting can cause stoppage. See Chapter 3 for repair procedures.

C. The CRUTCH:

The function of the crutch is to impart a slight push to the pendulum near the end of each swing.

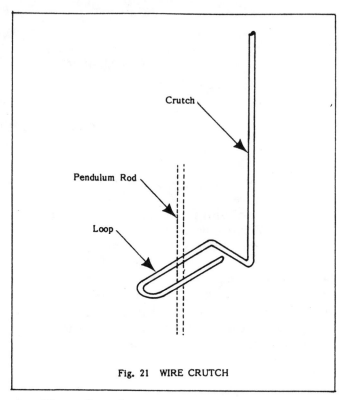

Fig. 21 WIRE CRUTCH

1. Wire Crutch:

The crutch on movements with wire pendulum rods is usually made of brass wire and is attached to the anchor, or to the arbor on which the anchor of the escapement is mounted. This wire is bent so that it hangs vertically, in front of the front plate of the movement and extends downward for several inches. At its lower end, it is bent to form an elongated rectangular loop, through which the pendulum rod passes. The loop should be horizontal and truly perpendicular to the movement.

There should be a little clearance between the pendulum rod and the inside of the crutch loop. The amount of clearance varies with different types of movements. If the sides of the loop are parallel, it is probable that the spacing is correct.

Many crutch wires are tapered so that they are larger in diameter at the top than at the bottom, where the loop is formed. It is not uncommon to find that this thin wire has been bent outward so that the loop is wider than it should be. The sides of the loop should be about parallel. If they are not, the side formed by the end of the wire should be carefully bent in, or out, using long nose pliers, to correct the condition. Recheck to make sure the loop is horizontal and the sides are perpendicular to the movement plate. Adjust as necessary.

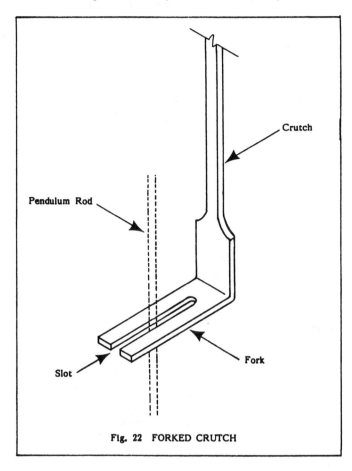

Fig. 22 FORKED CRUTCH

2. Forked Crutch:

Many pendulum rods are made of round steel or rectangular brass, both of which are much stiffer than the wire type discussed above, and not as subject to bending.

When such rods are used, the crutch is usually made of flat brass with a slotted fork at the bottom of the crutch. The slot is slightly wider than the width, or diameter, of the pendulum rod. The sides of the slot should be parallel. If they are not, and the ends are spread wider than the base of the slot, carefully squeeze the open ends toward each other,

using pliers. If the ends of the crutch slot are too close to each other, use a small screw driver to pry them apart. As with the wire crutch, the fork should be horizontal and perpendicular to the movement.

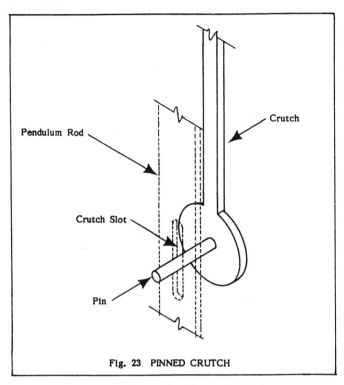

Fig. 23 PINNED CRUTCH

3. Pinned Crutch with Slotted Rod:

These rods may be made of any material, but are usually of wood or brass. They all have a slot to fit the pin on the crutch.

Crutches used with these rods are usually made of flat brass. At the lower end, they have either a pin or a formed metal end positioned at right angles to the crutch arm that fits into a slot in the rod. The pin should be perpendicular to the to the face of the movement, both horizontally and vertically. If it is not, correct by carefully bending the crutch, not the pin.

From top to bottom, the arm of the crutch should usually be roughly parallel to the face of the front plate of the movement, but in some cases is angled outward. It is important that the point of contact of the slot in the pendulum be near the center of the pin. Bend the crutch, if necessary to achieve this condition.

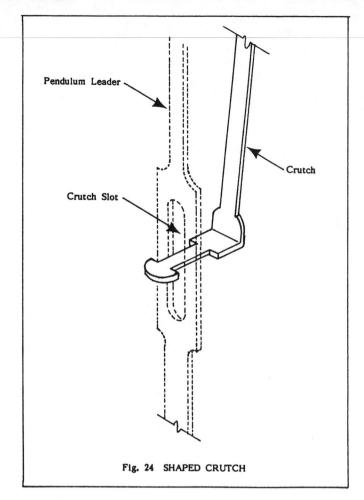

Fig. 24 SHAPED CRUTCH

4. Shaped Crutch:

Many movements of recent manufacture have composite pendulums with a slotted rod leader. The flat brass crutch for this type of pendulum functions in a manner similar to those with pins, in that it fits in a slot in the pendulum leader. The end of the crutch is larger than the part that fits in the leader slot, to prevent accidental disengagement of the leader from the crutch. The leader can only be disengaged by disconnecting it from the suspension spring and then rotating it 90 degrees so it clears the wide section at the end of the crutch.

It is common for these crutches to be bent at an angle away from the plate of the movement, from top to bottom, as shown in Fig. 24.

5. Centering the Crutch:

Make sure the clock is level.

When the pendulum rod is in place,

through the loop or slot in the crutch, or with the pin or crutch tab in the slot of the pendulum, attach the bob. The pendulum should hang so that it is about in the center of the crutch loop, front to back, or the pin or tab centered in the pendulum slot. It should be in contact with one side of the loop or slot. If it is not and, particularly, if it comes in contact with the crutch arm, the front or back of the loop, or slot, or is out of the slot, it must be adjusted.

With long nose pliers, grasp the top of the crutch, just below the point where it bends down, and bend it toward or away from the movement, until the pendulum rod is centered in the loop, or the pin or tab is centered in the slot.

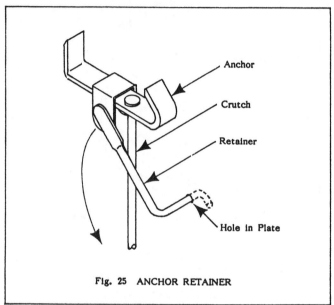

Fig. 25 ANCHOR RETAINER

D. The ANCHOR (Verge) RETAINER:

On some movements, particularly those that are weight-driven, the escape wheel and anchor are outside the front plate of the movement. The anchor, with its pallets, is attached by a rivet to a "U"-shaped bracket with holes in its ends. The holes in this bracket pivot on a pin mounted on the plate of the movement. The anchor is held in place by an "L"-shaped wire spring, one end of which is flattened and bears on the end of the pin. The other end goes through a hole in the front plate and is bent sharply behind the plate. It must be rotated away from the end of the support pin to remove the crutch assembly.

A frequent cause of stoppage of this type of movement is that the retainer spring slips off the end of the pin and comes into contact with the anchor, preventing movement.

Check to make sure that the spring retainer is properly positioned so that the flat end is centered over the pin and that it does not put pressure on the sides of the bracket or anchor.

Set the pendulum in motion. You may have corrected the problem.

E. HOW the ESCAPEMENT WORKS:

Before adjusting the beat, you should understand how the escapement mechanism functions, when properly adjusted. Shown here is a simple escapement, consisting of an escape wheel and a bent steel anchor, the ends of which form the pallets. The pallets alternately engage and release the teeth of the escape wheel as the anchor rocks back and forth on the pivot pin.

Springs or weights transmit power through the wheels and pinions of the movement time train to the escape wheel. Each time a tooth strikes a pallet, force is exerted to push the anchor in the opposite direction. This force is transmitted, through the crutch, to the pendulum rod, imparting enough power to keep it swinging.

The pin on which the anchor rotates should be positioned so that one of the pallets will instantly engage a tooth, as the other pallet releases the tooth with which it was engaged. If this pin is too close to the escape wheel, the pallets cannot disengage from the teeth. If it is too far away, the second pallet will not instantly engage the next tooth and the wheel will turn rapidly, until the pallet moves close enough to the wheel to touch the tooth tips. Since high speed is reached quickly, with significant inertia, damage to tooth tips will result. If your movement exhibits this problem, a means of correction will be found in Chapter 8.

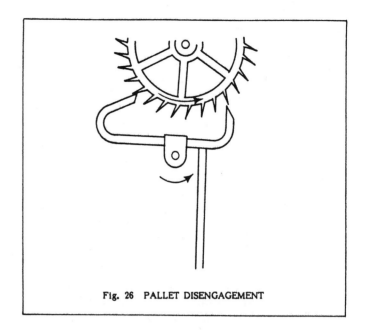

Fig. 26 PALLET DISENGAGEMENT

As the crutch and pendulum move to the right, the left pallet is about to release the tooth with which it has been engaged. The right pallet has cleared the top of a tooth.

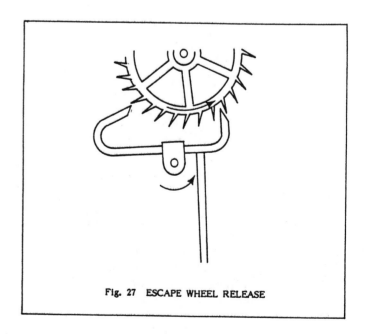

Fig. 27 ESCAPE WHEEL RELEASE

The left pallet has just cleared the tip of its tooth and the right pallet is between teeth. The escape wheel will now rotate until it is stopped by the right pallet engaging its tooth.

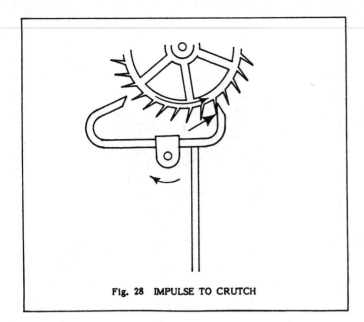

Fig. 28 IMPULSE TO CRUTCH

The right pallet has come in contact with a tooth, which now pushes against it and provides an impulse to the pendulum as it swings to the left.

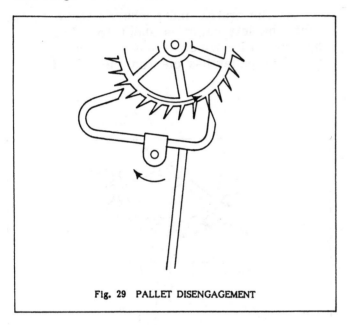

Fig. 29 PALLET DISENGAGEMENT

As the pendulum nears the end of its swing to the left, it is now the turn of the right pallet to release its tooth. The left pallet is now between teeth and ready to be impacted by another tooth, to add an impulse to the right.

F. TESTING and ADJUSTING:

Wind the clock and set the pendulum in motion. Can you hear it tick? If not, with the pendulum at rest, look at the anchor mounted on the arbor to which the crutch is attached. Observe the relationship of the tip of each pallet (arm) of the anchor to the tooth nearest it on the escape wheel. Is one pallet near the bottom of the space between the teeth, while the other is near the tip of a tooth? The position of each pallet should be nearly even, with respect to the teeth.

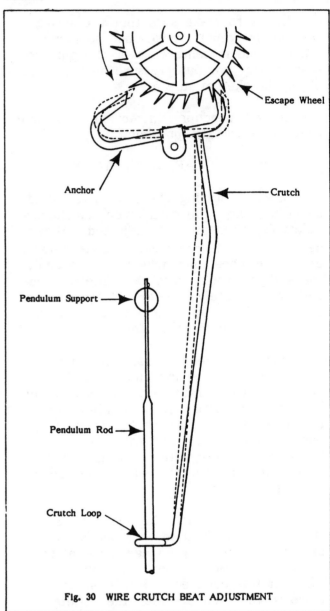

Fig. 30 WIRE CRUTCH BEAT ADJUSTMENT

1. Adjusting Pallet Engagement:

We need to change the relationship between the loop or fork on the crutch and the position of the pallets.

a. Wire Crutch:

In Fig. 30, the solid lines show

improper engagement of the pallets with the teeth, when the pendulum is at rest. The dotted lines show proper positioning of the pallets achieved by bending the crutch wire.

Grasp the crutch near the point where it is attached to the anchor. Pushing with a fingertip of the other hand near the loop, bend the crutch in the direction of the pallet that was near the tooth tip. Do this a little at a time and test after each adjustment, reversing the direction of bending if you have gone too far, until the clock ticks evenly. When you hear an even tick-tick (in beat), not a tick-tock (out of beat), the clock should run properly.

If there is no tick, slowly rock the crutch back and forth, observing the engagement of the pallets with the teeth. As one tooth is released, the escape wheel should move quickly to bring a tooth against the other pallet. If this doesn't happen, or if the action is sluggish, inadequate power is being transmitted to the escape wheel and the problem lies elsewhere.

Check to see whether one of the pallets is coming in contact with the top of a tooth so that it cannot drop into the space between teeth. If this is the case, it indicates that either a tooth has been bent, the pallets are excessively worn, pivot holes are badly worn, or the anchor arbor is out of position. If any of these conditions exist, it will be necessary to remove the movement from the case before proceeding further. See Chapter 8.

b. Bent Escape Wheel Tooth:

On rare occasions, one tooth of the escape wheel may have been severely bent, so that it prevents movement of the wheel. When a tooth is bent, its tip is moved out of position, reducing the space between it and the next tooth tip. Instead of dropping between teeth, as it should, the pallet hits the side of the bent tooth and the opposite pallet is prevented from raising enough to clear the tooth with which it has been engaged. The clock stops. See Chapter 8 for repair procedures.

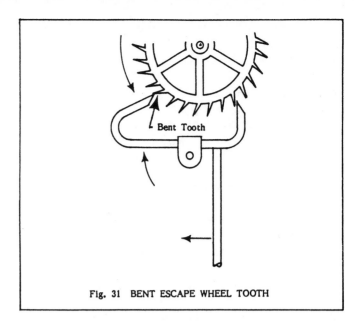

Fig. 31 BENT ESCAPE WHEEL TOOTH

c. Flat Brass Crutches:

Crutches formed from sheet brass vary considerably in width and thickness. The narrower ones may be bent, using care, in the same way we have described for wire crutches. In some cases, the crutch is attached to the arbor on which the anchor is mounted by means of a bushing and is secured by friction or a set screw. If a set screw is used, loosening it will let you adjust the crutch. If it is secured by friction, it may be possible to adjust it by holding the anchor firmly, while moving the crutch.

In some cases, the anchor is secured to the arbor by a set screw or by friction. Adjustment of the crutch position is made by rotating the anchor slightly on the arbor. If there is a set screw, loosen it before attempting adjustment.

To avoid damage to escape wheel teeth, great care must be exercised in making these adjustments. Do not use excessive force at any time.

As we will see in a moment, some flat crutches are provided with special devices that greatly simplify beat adjustment.

II. FRONT MOUNTED MOVEMENT:

Nearly all movements that are attached to the front of the clock case have pendulums supported on the back plate.

Many American movements of this type are housed in cases with a back that can be removed. Such backs are usually attached by screws. In some clocks of recent manufacture, the backboard fits into slots at the top and bottom of the case. To remove this type of board, lift it up slightly, which will bring the bottom out of its slot and pull the bottom outward. Taking the back off exposes the back of the movement and the pendulum.

Many clocks with front mounted movements have a large door which forms part or all of the back of the case.

Not infrequently, a coil gong, supported on an iron post is mounted on the bottom board, behind the movement. It will usually be necessary to remove the gong in order to check the action of the pendulum. The gong is held in place either by screws that run from the bottom of the case into the base of the gong support, or there is a threaded rod in the base of the gong support with a nut securing it to the bottom board. Loosen the screws or nut and remove the gong. It is frequently simpler to remove the base board with the gong in place, by removing the screws that hold it to the case.

With the pendulum and crutch accessible, follow the same procedures outlined in Section I of this Chapter to check the suspension spring, the crutch and the action of the escapement.

III. PENDULUMS SUPPORTED on the BACK OF THE CASE:

When the pendulum is supported separate from the movement on the back board of the case, as is typical of Vienna Regulators, there is usually provision for beat adjustment at the bottom of the crutch. There is most commonly a threaded rod, on which the crutch pin is mounted, with a knurled knob at one or both ends which, when turned, moves the crutch pin toward one side or the other, to change the relationship of the pendulum to the pallets. If the clock is out of beat, it can usually be corrected in this manner.

It is critical to make sure that the crutch pin is properly engaged in the slot of the pendulum, before making any adjustments.

A. ADJUSTING the BEAT:

1. Level the Case:

Using a level on a side of the case, adjust it so that the sides are vertical (plumb).

2. Wind the Time Train of the Movement:

It need not be fully wound. If yours is a weight-driven movement, simply be sure that the weight cord has at least two turns on its arbor. If your movement is spring driven, a couple of turns of the key will usually suffice.

3. Set the Pendulum in Motion:

Listen for an even tic-tic. If you hear an uneven beat, that is, a tic-tock, activate the beat adjusting device, a little in one direction. If the tock becomes more pronounced, move it in the opposite direction. Continue this procedure until you hear an even beat (tic-tic).

If there is no beat, try moving the adjusting device in either direction and set the pendulum in motion. Do this a little at a time until you have obtained a beat, or until you have reached the limit of the device's motion in that direction. If there is still no beat, move it in the opposite direction until a beat is obtained. Make fine adjustments until you have an even beat. If you reach the end of movement in each direction and still have no beat, it will be necessary to take the movement out of the case for further examination. See Chapters 7 and 8.

IV. BEAT ADJUSTING DEVICES:

A number of devices have been developed to vary the relationship of the pendulum to the pallets, in order to obtain an even beat.

A. THREADED PIN BEAT ADJUSTMENT:

Commonly found on Vienna Regulator and

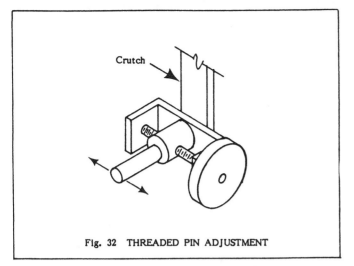

Fig. 32   THREADED PIN ADJUSTMENT

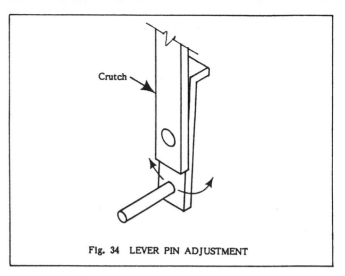

Fig. 34   LEVER PIN ADJUSTMENT

other wall clocks, this device may be part of or attached to the bottom end of the crutch.   The crutch pin is threaded onto a screw supported by a bracket.   A knob attached to one or both ends of the screw facilitate turning and move the pin to the right or left, depending on the direction of rotation.

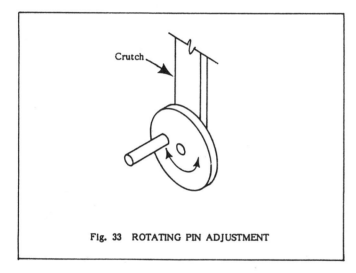

Fig. 33   ROTATING PIN ADJUSTMENT

B.   ROTATING PIN BEAT ADJUSTMENT:

This device is made up of a brass disc on which the crutch pin is mounted, off-center.   It is attached to the end of the crutch by a rivet at its center, which is sufficiently tight to hold it in position, but loose enough to be turned by the fingers to change the position of the pin. Rotating the disc moves the pin either to the right or to the left.

C.   LEVER PIN BEAT ADJUSTMENT:

Similar in function to the Rotating Pin device, this one has a pivoted lever

on which the crutch pin is mounted.   An ear on the back side of the lever facilitates rotation to adjust the pin position.

IV.   OTHER MOVEMENTS NOT OBSERVABLE IN THE CASE:

A.   ROUND MOVEMENTS:

Some movements are round in shape and are held in place in the case by long screws that connect the front bezel, on which the movement is mounted, to the back bezel, which usually has a hinged cover. The opening in the back is so small that reasonable observation of the movement is difficult, except to see whether or not the pendulum rod is properly connected to the suspension spring.

1.   Beat Adjustment:

Movements mounted this way depend on friction between the bezels and the case to hold them in position.   Beat adjustment can be accomplished by loosening the mounting screws and slightly rotating the bezels, until proper beat is obtained, then tightening the screws.

B.   OTHER MOVEMENTS:

There are other types of movements which must be removed before they can be closely observed and worked on.   You can easily determine this by careful observation, on a clock-by-clock basis.

Chapter 6

BASIC TOOLS for CLOCK REPAIR

It is an old axiom that a craftsman can be no better than his tools. In working with clocks, made largely of soft brass, this is particularly true. If a tool slips, at the very least it is likely to cause marring, while at worst, serious damage can result.

Only a few simple tools are necessary for doing the work we will discuss in this book. but they must be the right ones and in good condition. While the basic tools required are screwdrivers, pliers and a hammer, and tools by these names are available in most households, they may not be suitable for work on clocks.

For a very small investment, you can equip yourself with the proper tools at the beginning. Check the ones you now have against the specific requirements outlined in this chapter.

You won't have to buy every one of these tools before you start, but be aware of them and make sure you have the tool you need for the job you want to do.

I. SCREW DRIVERS:

Screws in common use are of two types, FLAT or Slotted head and PHILLIPS or Cross-slotted head. Screw drivers for each type are available in many sizes, from very tiny for watch and fine instrument work, to very large for use on heavy machinery. Blades of the better ones are made of tool steel and handles are commonly of plastic or wood.

At the time of this writing, very good standard screw drivers of both types, in the sizes required for our work, are available at a price of under $4.00 each at Sears stores and many hardware stores.

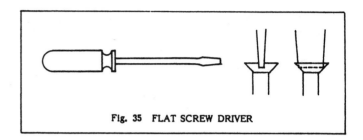

Fig. 35 FLAT SCREW DRIVER

A.   Flat Screw Drivers:

These screw drivers have a working tip which is tapered very slightly to its point. The point is square, flat across its face, with sharp edges. Size is determined by the width of the point and the length of the blade, measured from the face of the handle. Thickness of the blade increases with the width, to match the narrower slots in smaller screws and wider slots in larger ones.

The slots in screws vary in width from narrow in the smaller sizes, to wide in the larger sizes. It is most important that the tip of the blade fit closely into the slot of the screw and that it be sharp, so that the tip goes to the bottom of the screw slot.

If the blade is much thinner than the width of the slot of the screw, force will be exerted at a sharp angle near the center of the slot, reducing leverage and making it likely that the blade will slip out of the slot while you are turning it. Damage to both the screw head and surrounding parts of the movement or case is likely to result. It is important to use a screw driver of the proper size.

If the edges of the point of the screw driver have been damaged, or worn so that they are rounded, the point will not bottom in the slot and, when pressure is applied to turn the screw, it is almost inevitable that the blade will jump out of the slot. Screw driver blades can be sharpened by placing a piece of sand paper on a flat surface and moving the tip back and forth on it, while holding the blade vertical.

Short-blade screw drivers are easier to use when there is plenty of clearance around the screw to be driven or removed. In some cases, longer blades of a given size are necessary in order to reach the screw.

1. Recommended Flat Screw Drivers:

Tip Width	Blade Length
1/8"	4"
3/16"	4"
1/4"	6"
5/16"	8"

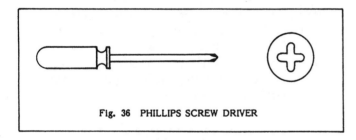

Fig. 36 PHILLIPS SCREW DRIVER

B. PHILLIPS SCREW DRIVERS:

If you are working strictly with older clocks, you are not likely to need this type of screw driver, since Phillips screws have only been used in clocks in comparatively recent times. However, when you do come across such a screw, it is essential to use the proper screw driver.

Phillips screw drivers have a V-point tip which is milled to form a cross, designed to fit closely into a similarly shaped recess in the screw head. The screw driver tip must exactly match the recess in the screw head. Too large a driver will not enter the slots, while one which is smaller than it should be may partially engage, but is likely to cause damage to the edges of the cross in the screw head.

The various sizes of Phillips screw driver tips are designated by number, from #0, the smallest, to #4, the largest. In some sizes, they are available in various lengths. Two sizes will handle most clock work.

1. Recommended Phillips Screw Drivers:

Number	Length
1	3"
2	4"

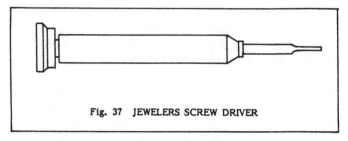

Fig. 37 JEWELERS SCREW DRIVER

C. JEWELERS SCREW DRIVERS:

Common screw drivers have wood or plastic handles. Also available and very handy for working with small screws, are Jewelers screw drivers, which have a very short blade mounted in a round metal body. The end of the body has a cap with a concave depression in it, which is sometimes free to rotate. In use, the index finger is used to apply pressure to the cap while the thumb and second finger rotate the body.

Usable Jewelers drivers have recently been available in large drug stores in sets of six, sizes #1 through #6 (1/32" to 9/64") in a plastic case, for under $2.00.

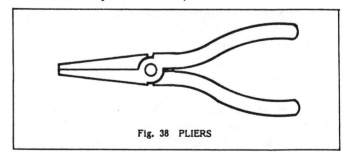

Fig. 38 PLIERS

II. PLIERS:

One of the tools commonly found around the house, pliers are available in an almost infinite number of styles, each with its own particular use. The slip-joint all-purpose style is fine for odd jobs around the house, but is totally unsuitable for use on clocks, because it is loosely jointed and too cumbersome for delicate work with small parts.

A. BASIC REQUIREMENTS of PLIERS for CLOCK REPAIR:

Regardless of the style, when buying pliers, it is a good idea to examine them carefully, checking for the following.

1. Pliers should be made of forged steel.

2. They must be of the solid joint type. This means that the faces of the two halves at the joint are machined to provide a close fit between themselves and with the pin that acts as a pivot. Action should be snug, but smooth, with little or no play at the joint.

3. When closed, the two faces of the jaws should be in intimate contact with each other. There should be no gap at the tip. If there are serrations (grooves) cut in the faces of the jaws, they should mate when the jaws are brought together.

4. Edges of the jaws should be fairly sharp, so they will grip small items securely.

B. RECOMMENDED PLIERS:

Each style of pliers is made in several sizes, determined by their overall length.

For clock work, it is not necessary to buy the most expensive pliers, as long as they meet the standards discussed above.

1. Long Nose 6" Pliers:

These pliers have a tapered and rounded nose which is about 1/8" wide at the tip. The face of the jaws is usually serrated and most have a wire-cutting section near the joint, which is useful for cutting soft wire, but will be damaged if used for spring wire, since that wire is harder than the steel of the plier jaws.

2. Flat Nose (Lineman's) 6" Pliers:

These pliers are essentially flat and have a square tip about 5/16" wide. The jaws are in contact for about 1/4" back from the tip, with parallel serrations on each jaw. There is a rounded open section with larger serrations, designed for gripping round objects, followed by a wire-cutting section near the joint.

We suggested modification of flat nose pliers for use with taper pins in Chapter 1. (Fig. 4)

3. Flat Face Long Nose 5" Pliers:

Flat face pliers are similar to regular ones, except that they have no serrations on the face of the jaws. For this reason, they do not grip as securely, but they are less likely to mar the piece being held.

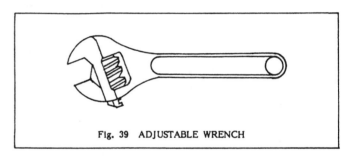

Fig. 39 ADJUSTABLE WRENCH

V. ADJUSTABLE WRENCH:

It is important to note that many of the nuts used on clock movements are made of soft brass. They have fine threads that are easily stripped if too much pressure is applied in tightening. While pliers may do the job, after a fashion, their jaws are not parallel and only the corners of the nut are gripped, so they can easily slip and cause damage.

We recommend the use of a very small, 4", adjustable wrench and carefully adjusting it to closely fit opposite faces of the nut. When tightening, use only enough force so that the nut can not be turned with the fingers. Threads are easily stripped when too much force is applied.

IV. HAMMER:

Talking about hammers in connection with clock movement repair sounds incongruous, but, as with other tools, the right hammer is an important implement. Ordinary claw hammers, the kind usually found in the home tool kit, usually weigh well over a pound and are much too large and heavy to even think of using on clock movements.

A small ball-peen hammer is useful in clock work and should be among your tools.

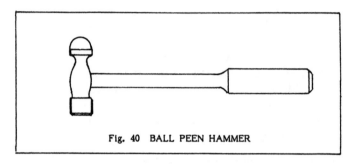

Fig. 40 BALL PEEN HAMMER

A. RECOMMENDED BALL-PEEN HAMMER:

A ball-peen hammer has a conventional flat surface on one end of the head, while the other forms a smooth ball. As its name indicates, it is used for peening, changing the shape of metal by repeated concentrated blows, as in peening over the tail of a rivet. For our purposes, a machinist quality hammer is not necessary and satisfactory ones can sometimes be found in the bargain baskets at local hardware or discount stores. We suggest:

Weight	Head Length
4 oz.	2 3/4"

V. SCREW and NUT RETRIEVER:

This is a simple device consisting of a metal tube with a spring-loaded wire running through it, a flat knob at one end and three or four bent wire prongs at the other. When the knob is depressed, the prongs are forced out of the tube and spread open so they can fit over the head of a screw. When released, they are pulled back and grip the head. There are two styles, one has a short, solid tube and fine wire jaws for small screws. The

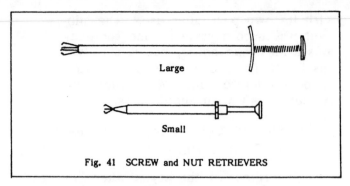

Fig. 41 SCREW and NUT RETRIEVERS

other is heavier and longer, with a flexible tube and is suitable for larger screws. They are lifesavers for both retrieving and starting screws and nuts in hard to get at places.

Both styles are inexpensive and should be a basic part of your clock tool kit.

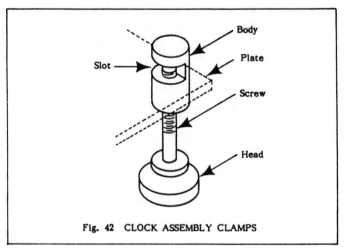

Fig. 42 CLOCK ASSEMBLY CLAMPS

VI. CLOCK ASSEMBLY CLAMPS:

These inexpensive little devices are not essential until you decide that you want to disassemble and reassemble a clock movement. When that time comes, you will find them most helpful.

Assembly clamps are available from most clock parts supply houses at modest cost.

Usually sold in sets of three or four, these clamps attach to the back plate of the movement to support it horizontally above the work surface, so that projecting elements are clear.

An assembly clamp consists of a threaded screw with a large round head which threads into a round threaded sleeve with a slot cut to the back of the threaded hole. This slot is slipped over the edge of the back plate, then the screw is tightened to secure it to the plate. The head of the screw acts as a foot to support the plate. Three clamps give good support for most movements.

VII. MEASURING DEVICES:

If you progress very far with clock repair, you will find a need for a means of measuring with a fair degree of accuracy.

A. LINEAR MEASURE:

1. Scales (Rulers):

Metric units of measure are common in clockwork and are not at all difficult to learn. We suggest that you have a good 6" steel scale, calibrated in both centimeters and inches with subdivisions in millimeters and 1/64".

2. Sliding Gauge (Caliper):

This is a device that allows you to measure inside, outside and depth dimensions by means of jaws that open and close, indicating measurement in inches or millimeters. Here again, we urge you to use the millimeter (metric) type.

B. THICKNESS MEASURE:

Particularly when ordering replacement mainsprings, it is important to know the exact thickness of the original.

1. Micrometers:

These devices are designed to make possible very accurate measurement of small dimensions. They are available with readings in thousandths of an inch or in 1/100 of a millimeter. We recommend a 0-10 mm (millimeter) metric micorometer.

Some micrometers are of very high precision and are correspondingly expensive. For clock work, somewhat less precise types are quite sufficient. At this writing, clock supply houses list satisfactory ones at under $20.00.

VIII. VISES and OTHER HOLDING DEVICES:

You may have a vise on your workbench now. Regardless of its size or style, it will probably be satisfactory for your clock repair needs for some time. When the occasion arises, however, you may want to consider purchasing one that is more specifically suitable for clock work. Most such vises are considerably smaller than the ordinary bench vise and thus more more convenient to use when working with small clock parts.

A. CLAMP MOUNTED VISES:

This type is designed to be easily attached to or removed from a bench or table. It has the usual jaws and screw, but instead of a plate with holes for permanent attachment, it has a yoke on the bottom side that fits over the edge of the table or bench like a clamp and is secured by a long screw.

B. SUCTION BASE VISES:

These small vises have a rather large rubber pad base attached to an eccentric lever which, when rotated, pulls the middle of the pad upward to create a suction, or vacuum, sufficient to hold the vise in place on a smooth surface, such as formica.

C. UNIVERSAL JOINT VISES:

Both of the vise styles discussed above, as well as the permanent mount type, are available in the conventional fixed position type and with a ball joint arrangement that allows you to position the jaws at various angles for maximum convenience. Some have nylon jaw faces that are ideal for holding brass parts, without marring them.

D. ROTATING JOINT VISES:

A little less flexible than the universal joint type, rotating jaw vises permit rotation of the jaws in one plane, for changing the angle of the face of the vise jaws.

E. HAND VISES:

These are small devices which are designed to be held in the hand while securely gripping small parts.

1. Conventional Jaw Type:

These are simply miniature standard vises, to which a handle is attached so that, after clamping the part in the jaws, they can be held in one hand while the other performs necessary work.

2. Pin Vises:

Pin vises are intended to grip small round pieces, such as the arbors of clocks or watches. They are made up of a metal tube with a gripping collet at one end which is tightened by a screw at the opposite end. The collet is a round steel piece with a hole at one end across which several narrow slots are cut to form jaws. The outside is tapered, so that when the collet is drawn into the body by a screw at the opposite end, the jaws are squeezed to grip the work piece. The diameter of the hole in the collet must be close to that of the piece to be gripped, since there is little movement when the jaws are tightened.

IX. CUTTING TOOLS:

CAUTION: Cutting tools remove material which may be difficult or impossible to replace. Before using them, be very sure that what you are going to do is necessary and will not further damage the movement.

A. JEWELERS FILES:

These are very small, very finely cut, files designed for delicate work. They come in a wide variety of shapes, which are shown in the catalogs of clock parts supply houses. Individually, they are inexpensive and, as you progress with clock repair, you may want to consider acquiring a few.

Before using files on clock parts, we strongly recommend that you practice on scrap material.

Proper use of files is to apply slight pressure only on the cutting stroke, that is, with a pushing motion. Lift the file up off the work on the return stroke.

B. JEWELERS SAWS:

Blades of jewelers saws are like those of coping saws, but very much smaller and with extremely fine teeth. They are held in a miniature coping saw frame.

Only when you have advanced to the point where you want to make replacement parts, or replace missing wheel teeth, are you likely to need a jewelers saw.

Chapter 7

TAKING THE MOVEMENT OUT OF THE CASE

With few exceptions, clock movements are supported within a case of some kind. When repairs that can be made with the movement in the case have not succeeded in making the clock run, as we discussed in Chapter V, the next step is to take it out of the case so that the search for causes of the problem can be continued.

There are five basic ways in which movements are supported:

1. Back Mounted: Attached to the back of the case.

2. Front Mounted: Attached to the front of the case.

3. Seat Board Mounted: Attached to a piece of wood, or metal, which is then supported by the case.

4. Front Plate Mounted: (Wood movements only.) Attached to vertical wood strips in the case by pins into the edges of the front plate.

5. Dial Mounted: Attached to the dial, which is then attached to the case.

Details of the mechanics of attachment can vary widely. We will cover basic principles of each method, with specifics of some of the styles most likely to be found in your collection.

CAUTION:

Before starting to take a movement out of its case, study the details of how it is attached carefully, following the guidelines given and determine the steps you will use in removing it.

Remove the pendulum rod, bob, and weights, if present.

Use the proper tools. Make certain screwdrivers fit the slots in screws and are sharp and square.

Use care and a minimum of force in every procedure, to avoid inflicting damage on the movement or case.

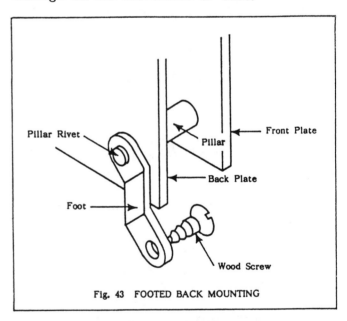

Fig. 43 FOOTED BACK MOUNTING

I. BACK MOUNTED MOVEMENTS - FOOTED TYPE:

This is one of the most common types of mounting. It is found on shelf clocks, wall clocks and even a few tall clocks.

Because, in most cases, the pivots of a movement extend beyond the face of the back plate, some means must be provided to prevent the pivots from coming in contact with the back board of the case, which would result in unacceptable friction. In a few cases, this was done by carving recesses in the backboard at pivot locations, so that the movement back plate could be screwed directly to the back board. An alternate method employs spacer washers between the back plate and the back board. Such movements have comparatively large holes in the front plate, through which a screwdriver may be inserted to reach the screws.

Usually, however, formed sheet steel feet were attached to the back of the movement on an extension of the pillars used to connect the front and back plates. These feet are formed so that they support the movement away from the back board. Short, rather large screws are driven, through holes in the feet, into the back board.

A. REMOVING FOOTED BACK MOUNTED MOVEMENT:

1. Remove hands. (See Chapter 1)

2. Remove dial. (See Chapter 2)

3. Remove pendulum bob and pendulum rod. (See Chapter 3)

4. Using a folded towel, or other soft cushioning material to protect its back, place the clock, back down, on a table or bench.

5. If the main springs of the movement are unwound so that their outer coils are in contact with, or very close to the sides of the case, wind them until they are well clear. To be sure the movement can be removed with a minimum of interference, it is a good idea to wind them until the outer coils are inside the movement plates.

6. Locate the screws holding the movement to the back board and select a screwdriver that closely fits the slots in their heads. You may want to use a flashlight, or other concentrated light source here.

7. Center the screwdriver in the slot of the first screw and, using only the force necessary, unscrew it.

8. If you have a screw retriever, remove the screw and place it in a small container. If you do not have a screw retriever leave the screw loosely in place after it is completely free, to be lifted out with the movement.

9. Repeat this procedure until all mounting screws are removed, or are loose in their holes.

10. Carefully grasp the sides of the front plate of the movement and lift it from the case. Try to keep the loose screws in their holes as you lift.

11. Especially on older clocks, oversize screws may have replaced one or more loose original ones. It is a good idea to replace the screws in the same holes in the backboard from which they were taken. Place them on the bench in a pattern corresponding to the holes in the back plate.

12. Place the movement, back down, on the table or bench.

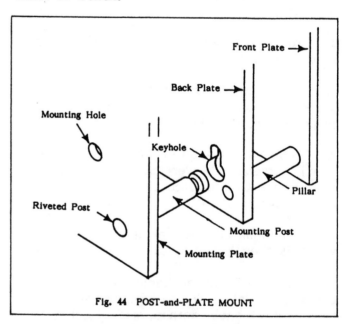

Fig. 44 POST-and-PLATE MOUNT

II. POST-and-PLATE BACK MOUNTED MOVEMENTS:

When pendulums are mounted behind the movement, space must be provided between the back plate and the case back board. In some Vienna Regulator clocks and a few American wall clocks, this is done by attaching a plate, on which posts are mounted, to the backboard. Keyhole shaped slots in the backplate of the movement fit into grooves at the front ends of the supporting posts, to lock it in place.

A. PROCEDURE for REMOVING POST-and-PLATE MOUNTED MOVEMENTS:

1. Grasp the movement and lift it up slightly, until the ends of the posts are in the large section of the keyhole openings in the backplate, then pull the movement forward.

III. FRONT MOUNTED MOVEMENTS:

Usually found in clocks with solid front cases, this type of mounting employs feet, similar to the back mounted type, (Fig. 43), but attached to the front plate, rather than the back plate.

Among older clocks, front mounting was used on almost all black mantel (rectangular case) clocks with wood or iron cases and on Tambour (Hump back, Camel back) mantel clocks. It is still being used today on similar styles.

Rectangular case clocks usually have a round opening in the case back, covered by a hinged or pivoted stamped tin door attached to the case back. This door may be opened so the movement can be seen, but is too small for the movement to pass through. The back board is fastened to the case by screws which must be removed and the back taken off, before the movement can be taken out of the case.

After taking the back off, check to see that the bob is on the pendulum and for other possible problems discussed in Chapter V. Their correction may enable the clock to run and make it unnecessary to remove the movement from the case.

Tambour clocks nearly always have a rather large hinged door on the back of the case, providing an opening large enough to permit you to reach and remove the movement.

Most clocks with front mounted movements strike the hours on a coiled wire gong, or on metal rods, mounted on

cast iron posts or brackets. These assemblies are attached to the bottom of the case by screws or a nut and bolt. They must usually be removed before the movement can be taken out of the case.

B. PROCEDURE for REMOVING FRONT MOUNTED MOVEMENTS:

1. Remove Hands. (See Chapter 1)

2. Remove Back Board, if there is not a large door in it. In older clocks, this is done by taking out screws that hold it to the case. In some more recently manufactured clocks, the back board may be held by pivoted tabs, which must be turned to allow the back to be removed. In others, the back simply slides into a deep groove in the top of the case and a shallower groove at the bottom. Lifting the board up into the top groove will allow you to pull it out at the bottom.

3. Remove the Bob from the Pendulum. In some cases, it may be necessary to do this through a slot provided in the bottom of the case. If so, move the case over the edge of the work surface enough so that you can reach up through the slot.

4. Using a Folded Towel, or some other cushioning material to protect the glass and the front of the case, carefully place the case face down on the work surface. Hold the front bezel in the closed position as you do so.

5. If Your Clock has a Solid Back with a Large Door, open the door.

6. Remove the Gong and/or Chime Rod Assemblies, by unscrewing the nut or screws that hold them to the bottom of the case. Use a small adjustable wrench. Support the assembly with one hand, so that it does not fall against the movement. When free, carefully remove it from the case.

7. Wind the Springs, so that the coils do not extend beyond the edges of the movement plate in such a way that they may interfere with the removal of the movement from the case.

8. Locate the Screws holding the movement

to the front board and select a screw driver that fits the slots in their heads. You may want to use a flashlight or other concentrated light source here.

9. Center the Screw Driver in the slot of the first screw and, using only the force necessary, unscrew it.

CAUTION: Especially with older clocks, it is not uncommon to find one or more oversize screws which have replaced loose originals. It is good practice to loosely reinsert screws in the same hole from which they were removed, after the movement has been taken out of the case.

10. If you have a screw retriever, use it to remove the screw. If you do not have a retriever, leave the screw loosely in place.

11. Repeat this procedure until all mounting screws are removed, or are loose in their holes in the feet.

12. Carefully grasp the sides of the back plate of the movement and lift it from the case. If the loose screws are still in the bracket holes, remove them and replace them in the holes in the case from which they came.

13. Because the hand shaft projects from the front of the case and the crutch from the back, the movement must be supported when you lay it down horizontally. If you have assembly clamps, attach them to the front plate, then place the movement on the work surface. If you do not have assembly clamps, support the front plate of the movement on the open end of a coffee can of suitable size to support the face of the front plate of the movement.

IV. SEAT BOARD and other BASE-MOUNTED MOVEMENTS:

This type of mounting is one of the earliest and has continued to the present. In their simplest form, early English and American clocks consisted of a movement, dial and hands, intended to be supported on a bracket or shelf provided with holes for the weight cords to pass through. To assure stability for this assembly, the

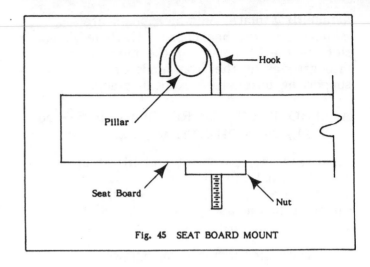

Fig. 45 SEAT BOARD MOUNT

plates of the movement were attached to a wood board, called a seat board, which was supported on brackets, or on a shelf.

On older clocks, the most common way of attaching the movement to the seat board is by means of two small rods, which have a hook at one end and are threaded at the other. The hooks fit over each of the two bottom pillars of the movement, pass through the seat board, then are secured with a nut. These are referred to as "seat board hooks".

When cases were added, for decorative and protective reasons, the seat board was continued and the shelf on which it rested was simply incorporated inside the case.

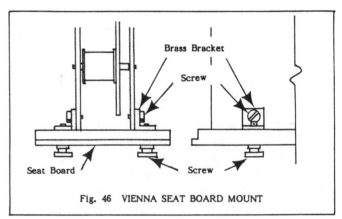

Fig. 46 VIENNA SEAT BOARD MOUNT

Some old Vienna Regulator clock movements were attached to a wood seat board by means of L-shaped brass brackets fastened to the plates of the movement by small screws. The foot of each bracket has a hole threaded to receive a long steel screw with a large brass head. This screw goes through the seat board and, when tightened, secures the movement to the board. Commonly, there is one such bracket at the center of the front plate and two at either side of the back plate of the movement. The board has rabbeted edges which fit into grooves in wood brackets attached to the back board. The movement is removed from the case by simply sliding it forward.

In a variation of this design, a metal plate was used in place of the wooden board and grooves were provided in metal brackets.

In some clocks, notably some Vienna Regulators and other clocks of more recent manufacture, holes are drilled vertically through the bottom pillars and threaded to receive a long bolt with a knurled head. Usually, when this is the case, there is no seat board and the plates of the movement are secured directly onto the top of a slotted metal bracket attached to the back of the case.

With seat board and other base-mounted movements, the dial is usually attached to the movement and the pendulum is at the back. The pendulum may be supported by a suspension spring on the back plate of the movement or it may be independently supported on the back board of the case. In either case, the movement crutch is at the back and a pin on the crutch fits in a slot in the pendulum. When the pendulum is independently supported, the movement can be withdrawn without removing the pendulum.

Some movements mounted in this way, with their attached dials, may be very heavy. It is important to anticipate this when removing them and be prepared to handle them safely, without damage to either the case or the movement, or injury to yourself.

Seat board and other base-mounted movements may be supported and secured in place in a variety of ways, as has been indicated. A sampling follows. If the clock you are working on is not exactly like any of the ones described, it will probably incorporate a combination of the basic elements touched on. Careful study of the examples discussed should make it possible for you to deal with it.

A. PROCEDURE for REMOVING BASE-MOUNTED MOVEMENTS:

1. Remove weights, if yours is a weight-driven clock.

2. If yours is a tall clock, with a removable hood, remove the hood by sliding it forward.

3. If yours is a tall clock of recent manufacture with a fixed hood, removal of the movement may be difficult. It is suggested that you consult literature usually supplied with such clocks in order to determine the best method of removal. In some cases, there is a wood panel in the back of the hood giving access to the movement.

4. Determine if the pendulum is supported on the movement and, if it is, disconnect and remove it.

5. Examine the movement and its supporting structure carefully to determine how it is secured. See the discussion of specific types which follows.

6. Proceed to remove the movement from the case.

B. MOUNTING TYPES:

1. Wood Seat Boards - Tall Clocks:

These are usually rather heavy boards about 3/4" in thickness. Commonly, they simply rest on supports provided in the case, sometimes with stop blocks glued in place on the supporting structure for the purpose of positioning the seat board. If this is the case, the board and movement can simply be lifted out. It is possible that, originally, or at some later date, screws have been added. Look for them, if the movement does not come free easily.

In most cases, with support at the ends, the seat board will act as a base on the work surface.

2. Wood Seat Boards - American Brass-Movement Weight Driven Shelf Clocks:

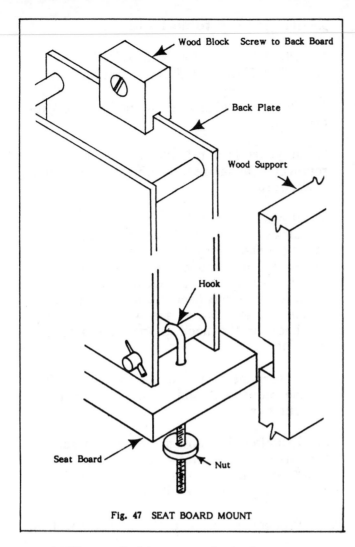

Wood Block   Screw to Back Board

Back Plate

Wood Support

Hook

Seat Board

Nut

Fig. 47 SEAT BOARD MOUNT

Millions of this type of clock were made in the United States during the nineteenth century. Common case styles were half-column, ogee and column-and-cornice.

Cords for transmitting force from the weights to the movement run from the winding drums on the movement upwards and over pulleys set in elongated holes on each side of the top board of the case. Weights must be removed and the cords, with their hooks, passed back over the pulleys and into the case, so that the movement can be taken out of the case.

Two relatively thin boards, perpendicular to the back of the case and usually fixed between the top and bottom boards of the case, provide a sort of housing for the weights. A groove is cut on the inside of each board, into which the movement seat board slides. When it is properly in position, the seat board is

in contact with the case back board. The movement is attached to the seat board with seat board hooks and is secured at the top of the back plate by a small block of wood with a slot fitting over the plate. This block is screwed to the back board. After the top block and weights have been removed and the cords run through the holes in the top of the case, the seat board and movement can be slid forward, clear of the supporting grooves.

The crutch is nearly always in front of the front plate on these movements, so they may be placed on their backs on the work surface, after removal from the case.

3. Vienna Regulator and Similar Base-
     Mounted Movements:

NOTE: Some Vienna movements are back-mounted. See paragraph II of this chapter.

Nearly all Vienna Regulator clocks have the pendulum supported on the backboard of the case, independent of the movement. Check this, however, and, if the pendulum is attached to it, don't attempt to remove the movement until the pendulum is disconnected.

a. Wood Seat Board:

Early Vienna Regulator clocks had wood seat boards to which the movement was attached by machine screws. In some instances, a brass angle bracket was mounted on each side at the bottom of the backplate of the movement and threaded to receive the mounting screw. Sometimes a hole was drilled in the bottom pillars and threaded to receive the mounting screw.

The seat board frequently was rabetted at its outer edges to form a tongue, about half its thickness, which fits into a groove cut into the support. Supports were usually wood brackets attached to the back of the case.

After removing the weights, the seat board can be moved forward, out of the groove.

b. Metal Seat Plate:

In one version of this type of

mounting, the wood board is replaced by a thinner brass plate that slides in a groove in metal brackets attached to the back board of the case. The plate is sometimes secured by a screw on each side. If your clock has such screws, and there is a metal plate, simply loosen them to withdraw the movement.

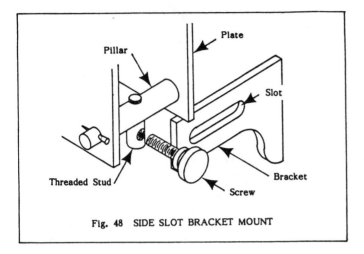

Fig. 48 SIDE SLOT BRACKET MOUNT

c. Side Slot Bracket Mount

In this mount, a short bracket or pin, with a threaded horizontal hole, is attached to the bottom of the seat plate, or to a pillar, so that it fits just inside the mounting bracket. There is a short slotted hole in the side of each mounting bracket, through which a brass-headed steel screw is inserted into the threaded hole in the bracket or pin. These screws must be completely removed to withdraw the movement.

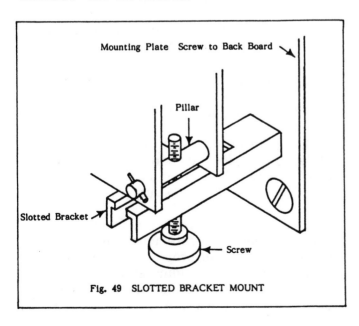

Fig. 49 SLOTTED BRACKET MOUNT

c. Slotted Bracket Mounting:

Commonly used on so-called "R-A" Vienna regulator spring driven clocks, this method employs support brackets, of inverted "U", or channel, cross section. A slot in the top is open at the front. The plates of the movement rest directly on the top of the bracket. Steel screws with large brass heads, so they bear on the bottom of the bracket flanges, are threaded into holes in the bottom pillars of the movement, to secure it in place.

Loosening the two screws will permit the movement to be pulled forward to remove it from the case.

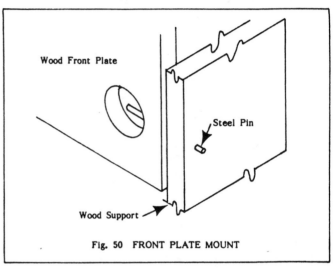

Fig. 50 FRONT PLATE MOUNT

V. FRONT PLATE MOUNTED MOVEMENTS:

American wood movement shelf clocks produced by mass-production methods in the early 19th century, were commonly supported by their front plates.

These are comparatively large movements with wood front and back plates. The case has two vertical wood strips, located so that the movement fits closely between them. Four large holes are located near the side edges of the front plate, just inside the four corner pillars. Small holes are drilled through the supporting strips and into the sides of the front plate of the movement, at the outside center of the large holes. A small steel retaining pin, passed through these holes, serves to position and support the movement.

The retaining pins are of such a length that their ends will be visible in the holes in the front plate and should extend slightly beyond the face of the support board.

A. PROCEDURE for REMOVING FRONT PLATE MOUNTED MOVEMENTS:

1. Remove hands and dial. (See Chapter 2)

2. Remove bob and pendulum rod. (See Chapter 3)

3. Remove weights.

4. Pass cords and weight support hooks through holes in top of case, to the inside of the case.

5. Locate the pin in one of the bottom holes of the front plate and, using a small screwdriver, push it toward the outside. Grasp the pin with the fingers and remove it. If a pin does not come out easily, it may be necessary to use pliers. When this is the case, apply force gently in the direction of the line of the pin, to avoid enlarging the hole or bending the pin.

6. Repeat step 4 with the other pins, supporting the movement with one hand until the last pin is removed and the movement is free.

7. Lay the movement on its back on the work surface.

VI. DIAL MOUNTED MOVEMENTS:

This is the most common form of mounting for typical round French movements and is found occasionally in other types of clocks. It can usually be identified quickly by the fact that there is a surface-mounted brass bezel on the front which carries another hinged bezel with glass and, on the back of the case, a bezel of similar size, with a round metal door. This back door is usually hinged, but may simply be inserted into the bezel and held by friction. Large holes in the front and back of the case are of a size to closely fit the outside of a flange on the inside of the bezels.

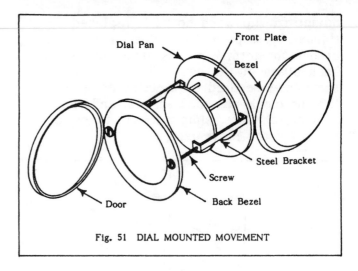

Fig. 51 DIAL MOUNTED MOVEMENT

The dial is fixed to the bezel and the movement, in turn, is attached to the dial. The pendulum is suspended from the back plate of the movement.

A narrow flat steel strap is attached on each side, by a rivet, to the the back of the front bezel. These straps extend toward the back of the case where they bend sharply inward to form an ear in which there is a threaded hole. A hole in each side of the back bezel lines up with the hole in each strap. A long steel screw passes through the hole in the bezel and is screwed into the hole in the ear. When it is tightened, this pulls the front and back bezels against the case and secures them by friction. In some instances, probably where the threads in the strap ears have been stripped, a nut is used, in back of the ear.

A. PROCEDURE for REMOVING DIAL MOUNTED MOVEMENTS:

1. Remove strike bell, if there is one.

2. Remove pendulum rod and bob. Because of the relatively small opening in the back of the case, if you have them, a pair of large tweezers, or forceps can be very helpful here. Use great care, so as not to damage the delicate suspension spring.

3. With one hand, hold the front bezel gently against the case front and, with a screwdriver of the proper size, loosen and remove the two retaining screws at the rear bezel. Remove and set aside the back bezel assembly, while maintaining pressure

on the front bezel.

4. With both hands, remove the front bezel assembly, with the dial and the movement, from the case by pulling it forward.

5. Hold the front bezel to prevent accidental opening and lay the movement, with the dial down, on the work surface.

Chapter 8

EXAMINING THE MOVEMENT

OUTSIDE THE CASE

We now have the movement where we can get our eyes, and our hands, on it and can take a much closer look, remembering that we are seeking to find why it does not run. Along the way, however, we may learn more about some of the basic functions of clock movements.

Since this is a primer, defined in the dictionary as a "simple book....giving the first principles of a subject", we will not be able to go into specific details of every movement type. However, it is hoped that we will cover many of the basic principles that apply to nearly all clock movements. As stated in the Introduction, we will limit suggested remedies to those which can be made without sophisticated tools, or extensive knowledge of the art and science of horology.

We will explain the working of the various parts of the movement, their relationship to each other and some of the most common faults that may be found. At this point, no disassembly of the movement is involved. That will be dealt with in the next Chapter.

WORKING WITH A MOVEMENT IN HAND

By far the best way to understand this step by step examination of a movement is to have one in hand to refer to as you read and compare with the figures. This can be any type of pendulum movement. While some of the details may not exactly match the movement you have, the principles common to all pendulum clocks will be readily observable.

I. GENERAL CONDITION:

A. CLEANLINESS:

1. Dirt:

Many old movements have spent much of their lives in far less than ideal conditions. It is not at all uncommon to find them encrusted with all sorts of dirt, and not infrequently with nests of mud-dauber wasps and other insects. It is obvious that dirt of any kind can cause enough friction to prevent the clock from running. Heavy dried mud deposits should be removed by the use of a smooth metal instrument, such as a table knife.

2. Excess Oil:

Most people think that if a clock does not run, it is because the movement needs oiling. They seldom realize that too much oil, or lubricant of the wrong kind, can cause much more trouble than it can cure. A great many clocks are found with their movements literally saturated with oil, which has attracted and held significant amounts of abrasive dust. Even worse is graphite. Over the years, this mixture of lubricant and dust tends to become gummy and can literally stop the movement.

For a great many years, there was a common belief that placing a small open container of coal oil (kerosene) in the bottom of a clock case, so that its vapors could lubricate the movement, was deemed essential to good operation. Old clocks with weight driven movements frequently still smell like kerosene. This practice resulted in the buildup of varnish-like coatings on all parts of the movement, holding dust and increasing wear. In pivot holes it can become so gummy as to act almost like an adhesive.

II. PRELIMINARY MOVEMENT CLEANING:

If you do not intend to fully disassemble and clean the movement, we suggest that you use a new dry 1/2" nylon paint brush to gently clear away as much of the dust and dirt as you can.

If the movement is VERY dirty, it may be necessary to give it a preliminary cleaning, in order to properly examine it and to make a reasonable assessment of what may be needed to repair it.

CAUTION: We will suggest a procedure for cursory preliminary cleaning. DO NOT USE IT UNLESS YOU ARE PREPARED TO FOLLOW UP WITH THOROUGH CLEANING as described in Chapter 10.

If you use this procedure and do not follow up with thorough cleaning after fully disassembling the movement, rust and serious damage can occur, particularly to the springs and pivots.

1. Suggested Procedure IF CLEANING IS NECESSARY:

a. Obtain a container, such as a plastic bucket, of such a size that the movement will fit into it, comfortably.

b. Make a solution of liquid household detergent and hot water, as directed by the manufacturer for general cleaning.

c. Immerse the movement in the solution and allow it to soak for about 15 minutes, sloshing it around occasionally in the solution.

d. Use a new 1/2" paint brush to gently brush all surfaces with plenty of solution.

e. Rinse in HOT running water to remove all detergent.

f. Gently shake the movement, holding it in several different positions, to remove as much water as you can.

g. Place the movement on a folded towel and, if you have one, use a hair dryer, set on high heat and maximum fan speed, to dry it. If you do not have a hair dryer, place the movement in a warm oven (about 220 degrees) for about 20 minutes. ALLOW TO COOL BEFORE HANDLING.

THOROUGH, QUICK DRYING is important to minimize the formation of rust, on steel parts, particularly pivots and springs.

III. EVIDENT DAMAGE or DEFICIENCY:

It is possible that the movement has suffered damage that will require skills beyond the scope of this book which can be detected at this point. When this is the case, you should proceed no further, but seek outside help.

A. LOOSE or MISSING PARTS:

It is not at all uncommon to find movements from which parts have been removed or lost.

1. Screws and Nuts:

You have already checked for missing suspension springs and would have been aware if the crutch and anchor escapement were missing. Now check to see if any screws required to hold the support for the suspension spring to the movement plate are missing or loose.

Are any nuts used to secure the front plate to the pillars loose or missing? Look for other threaded holes or studs from which screws or nuts may be missing.

Make sure all required nuts and screws are in place and secure. Don't over tighten. Brass nuts are easily stripped.

2. Arbor Assemblies:

Now look for missing arbor assemblies, an arbor (shaft) with its wheel and pinion, which may have been scavenged from your movement. Ordinary striking movements have two trains of wheels, running from the largest, on the arbor that carries the spring or cord drum, at the bottom, to the smallest, on the arbor that carries the escape wheel on the time side or the fly on the strike side, at the top. Each wheel should engage with a pinion on the succeeding arbor. If it does not, an arbor is missing.

If an arbor assembly is missing, a replacement from an identical movement will be necessary, since new parts of this type are not generally available.

Sometimes, a local fellow collector may have accumulated a stock of old movements from which he uses parts and might have what you are looking for. On occasion, at local NAWCC meetings and frequently at Regional meetings, members have old parts for sale. If you have joined NAWCC, take your movement to a meeting and look for what you need.

If you cannot find a replacement, a skilled clock maker with the proper equipment can make an entire replacement assembly. Be forewarned, this is a comparatively expensive operation.

B. LOOSE CONNECTION of PLATE to PILLARS:

Front and back plates of the movement are separated from each other by several round pillars or posts. They are usually permanently attached to one plate by riveting. The other plate has holes which fit over a short extension of the pillar and rest on a shoulder.

The pillar extension fitting through a hole in the plate, may be threaded, in which case a nut secures the plate. Or, the extension may have a small transverse hole drilled through it, so that its back side is just below the surface of the plate, when it is in position. A taper pin pressed into the hole exerts pressure against the plate and holds it secure, by friction.

If either nuts or taper pins are loose or missing, the plates are not held securely and may have separated enough to allow one or more pivots to come out of their holes.

C. BROKEN or DISPLACED PIVOTS:

Check for displaced or broken pivots by holding the movement in a vertical position and looking at it from the side, that is, between the plates. All of the arbors should be parallel. If they are, tighten the pillar nuts or replace and secure taper pins by pressing them securely home.

The pivots are the extensions at the ends of each arbor (shaft) that are smaller in diameter than the body of the arbor and fit into holes in the front and back plates of the movement to act as bearings. Normally, they extend a little beyond the outside edge of the hole in which they fit. Sometimes, pivots break off the arbor. When they do, the end of the arbor is unsupported.

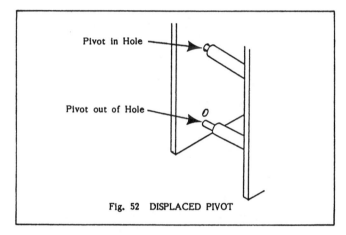

Fig. 52 DISPLACED PIVOT

If one or more arbors are not parallel, either a pivot has come out of its hole, or a pivot has broken and allowed the arbor to fall at that end. In rare cases an arbor may be bent or broken, usually the one carrying the fly governor.

Should any of these problems be found, correction will require disassembly of the movement. See Chapters 10 and 11.

1. Broken Pivots:

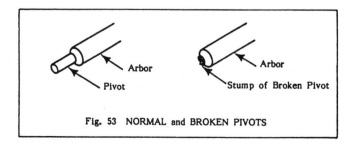

Fig. 53 NORMAL and BROKEN PIVOTS

If a pivot is broken off, or an arbor is bent or broken, the only alternatives for the novice are to replace the entire assembly of arbor, wheel and pinion with one from an identical old movement, or to have a new pivot installed by a clock maker equipped to do this kind of work.

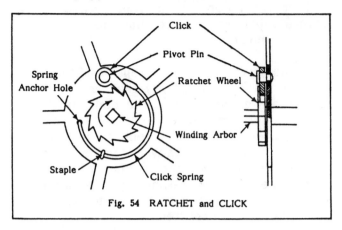

Fig. 54 RATCHET and CLICK

D.   CLICKS and CLICK SPRINGS:

In clock making, the term "click" refers to a pawl or detent that acts with a ratchet wheel to allow winding of the spring or cord, then locks the wheel and the arbor so that power can be transmitted to the movement.   The ratchet wheel is attached to the winding arbor, while the click, shaped like a comma with a hole in it, fits freely on a headed rivet secured to the first (great) wheel.   As the arbor and ratchet wheel are turned, the click is lifted up by each tooth of the ratchet and then forced down against the face of the next tooth, by a spring, called the click spring.

If, during winding, a spring arbor turns and meets with normal resistance of the mainspring, but does not lock when turning force is stopped, the click is not performing.   With many, but not all movements, it is possible to observe both the click and its spring without disassembling the movement.   If the click is not functioning, correction will require disassembly of the movement.   This will be discussed in Chapter 12.

1.   Clicks:

Clicks are nearly always made of brass and are attached to the wheel by means of a headed brass rivet passing through a hole in the thick end of the click.   The

rivet has a shoulder so that, when in place, there is only slight clearance between the head of the rivet and the face of the click.   This serves to confine the click and keep it in line with the teeth of the ratchet wheel.

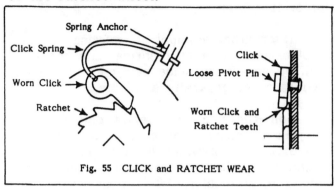

Fig. 55 CLICK and RATCHET WEAR

a.   Click Problems and Solutions:

Considerable work is done by the click during the working life of a clock and substantial force is exerted on the riveted connection of the pin in the wheel.   Over time, the rivet may become loose.   When it does, the click may fail to engage with the ratchet and become ineffective.   In addition, as the pin becomes progressively more loose, severe wear of the engaging tips of the ratchet teeth and of the click may occur, eventually causing the click to slide off of the teeth of the ratchet wheel.

When either or both of these conditions occur, the rivet and/or the click and ratchet must be replaced.   This requires disassembly of the movement and will be discussed in Chapter 12.

2.   Click Springs:

There are two basic forms of click spring:

a.   Attached to Click: (Fig.   55)

In this form, a piece of round spring wire is inserted into a hole or slot in the top, or convex side of the click and then riveted to hold it in place.   It is then curved so that it points in the direction of the click point and its end inserted under a tab raised up from a spoke of the great wheel.   This exerts a downward pressure on the tip of the click to keep it engaged with the ratchet wheel.

b. Attached to Wheel:

In this form, a round or flat spring is secured to the great wheel so that its end exerts pressure on the top of the click nose to maintain pressure on its engaging tip.

(1) Round Wire Click Spring: (Fig. 54)

When a round wire click spring is attached to the great wheel, a narrow groove is provided in the top of the click into which a flattened end of the spring fits, to keep it in place. All too often, this flat end comes out of the groove, so that no pressure is exerted on the click and it fails to engage the teeth of the ratchet and the spring cannot be wound. Usually, simply replacing the spring in the groove, using the tip if a small screw driver, corrects the problem.

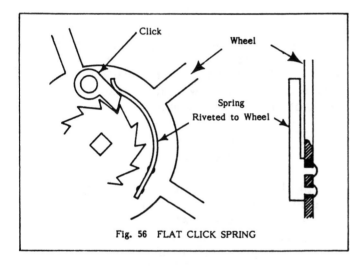

Fig. 56 FLAT CLICK SPRING

(2) Flat Wire Click Spring:

Flat click springs are curved to exert pressure and have a reverse curved section at the outer end which rests on the flat top of the click. The other end is formed with two short stubs that fit through holes in the great wheel and are securely riveted at the back of the wheel. These springs sometimes slip off of the click, so that no pressure is exerted and the click does not lock the ratchet. If restoring the click spring it to its proper position does not solve the problem, it will have to be dealt with after disassembly of the movement.

E. MAIN SPRINGS:

Main springs will be discussed in some detail here, because, having taken the movement out of the case, this is the first time we have been able to observe them closely.

The large springs which store power to drive the time, strike, or chime trains of a movement are called "Main Springs". This is to distinguish them from other springs found in the movement. They will vary in thickness, width and length, depending on the design power requirements of the movement.

Mainsprings are made of an alloy steel which is formed into a coil, then heat treated to make it resistant to bending. In its final form, spring steel is hard and brittle. At rest, that is, not restrained at either end, a mainspring will have considerable space between its outer coils, with the space between coils decreasing progressively to the innermost ones. In this condition, the spring is completely relaxed and stores no power.

Power is stored in the spring by winding, pulling the coils closer and closer together. This action is resisted by the spring, which wants to return to its relaxed state. It is this resistance which, when restrained by the ratchet and click, stores the power necessary to run the movement.

When the inner end of the main spring is attached to a winding arbor and the outer end is secured to a fixed point, turning the arbor draws the coils together, while the spring wants to stay relaxed. Thus resistance of the spring is overcome by the force exerted in winding and the storing of power results. It is the spring's constant effort to return to its original relaxed state that exerts a turning force on the arbor and on the train of wheels it powers.

The direction of winding is determined by the way the spring is installed in the movement. Turning it over reverses the direction. The proper direction is determined by the design of the movement.

As used in clock movements, the spring is never completely relaxed. It is always confined, at least to some degree and stores power which, if suddenly released, can cause damage to the movement or injury to the unwary repairman. Springs must always be treated with respect, even when mostly unwound..

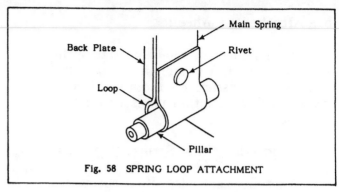

Fig. 58 SPRING LOOP ATTACHMENT

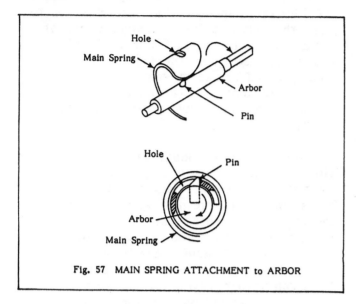

Fig. 57 MAIN SPRING ATTACHMENT to ARBOR

1.   Types of Mainsprings:

Main springs differ not only in dimension, but in the way in which they are attached. The inner end of the coil is always provided with a hole which fits over a short pin projecting from the winding arbor. As the arbor is turned (in the proper direction) the pin pulls on the spring, causing it to wrap around the arbor.

The outer end of the mainspring coil is a different matter. Two methods of attaching the outer end define the two basic types of mainsprings:

a.   Loop-End Main Springs:

In most clock movements, a small loop is made in the end of the spring, slightly larger in diameter than that of one of the lower pillars of the movement, over which it fits. Thus, the outer end of the coil is restrained when the spring is wound on the arbor, causing power to be stored in the spring.

Loop-end main springs are usually readily visible from both sides of the

movement, although there is occasionally a thin steel plate on one side of it.

(1) Spring Restraints -Loop End:

To prevent damage to other parts of the movement by force exerted by the outer coils of a mainspring as it unwinds, some movements have as open flat steel ring surrounding the mainspring. In other cases, a sturdy steel pin is riveted to one plate of the movement in a position to limit expansion of the outer coil.

When no such restraints are provided and the spring unwinds, the outer coils may exert considerable pressure on the inside of the case, which then acts to restrict further movement.

b.   Hole End (Enclosed Barrel) Main Springs:

In this arrangement, the mainspring is confined within a brass cylinder, formed as part of the great wheel. The spring, in a partially wound condition, is inserted into the inside of the barrel, with a hole in its end hooked over a steel pin attached to and projecting into the inside of the barrel. The inner coil of the spring has a hole which fits on a pin projecting from the winding arbor. A flat brass cover, snap fitting into a groove on

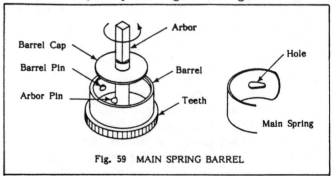

Fig. 59 MAIN SPRING BARREL

the inside of the barrel, on the side opposite the wheel, completely encloses the main spring.

The barrel obviously is good protection against damage to other parts of the movement in the event a spring breaks or a click fails. It further protects the spring from dust.

2. Main Spring Problems and Solutions:

a. Broken Springs:

It is a common misconception that overwinding can cause mainsprings to break. They are designed to be fully wound. If a regular clock key is used for winding, breakage is far more likely to result from flaws in the steel, or what engineers call metal fatigue. Failure can manifest itself in a number of ways, sometimes a single break at one point in the spring and sometimes by many breaks occurring throughout its length. The most common breaks are found near the holes used to secure the spring to the arbor pin, where the spring is weakest and the stresses most concentrated.

With loop end springs, failure is usually evident, to the eye, but sometimes, if the break occurs near the inside end, it may not be. If there is no resistance felt when the winding arbor is turned with the key, it is a sign that this may be the case, or it may be that the spring has become unhooked from the pin on the winding arbor. In either case, disassembly of the movement and full relaxation of the spring will be necessary to correct the problem.

When mainsprings are housed in barrels, it is not possible to see the condition of the spring. If winding is not effective, there is a problem the nature of which can only be determined after disassembly of the movement. We will go into that in Chapter 11.

b. Rusted or Dirty Main Springs:

If a mainspring is fully or even partially wound, but refuses to release its power, or releases it intermittently, with sudden uncoiling from time to time,

it is probable that it is rusty, or dirty. Unless all surfaces of the spring are clean and properly lubricated, they will tend to stick to each other, or to rub with such friction that power is not released continuously and smoothly. This condition is readily evident during winding, when uneven gaps can be observed between coils, or there is a sudden release of power with a snapping sound as individual coils break the bond and violently uncoil.

There is no quick satisfactory solution to this problem. Oiling may provide some temporary relief, but is unlikely to permanently solve the problem. Rust-solvent oils are likely to aggravate it by spreading rust to unaffected surfaces which, when the oily base evaporates, leave more widespread residue and increased friction.

In Chapter 11, we will discuss methods of cleaning and lubricating main springs.

F. WHEEL and PINION TEETH:

Damage to wheel or pinion teeth can result from wear, but more commonly is caused by the breaking of a mainspring or by abuse at the hands of an untutored repairman.

You should carefully examine each tooth of the movement. Look first for missing teeth. Look particularly in the area where wheel and pinion teeth are engaged. If one or more teeth are missing or seriously damaged, the point of engagement is where interference will occur and cause stopping of the movement.

If bent or broken teeth are found, disassembly will nearly always be necessary.

1. Pinion Types:

a. Solid Pinions:

Nearly always made of steel, except in movements made of wood, solid pinions are cut from a solid piece of metal, or wood and have teeth (leaves) shaped to provide fairly continuous engagement with those of the wheel with which they mate. While it is not uncommon for wood pinion leaves to be broken, this is less likely with solid steel pinions, although it does happen.

There can be significant evidence of wear on the surface of the pinion leaves without important effect on the function of the movement.

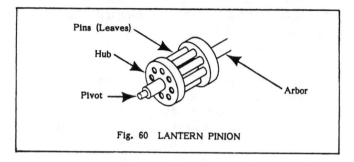

Fig. 60 LANTERN PINION

b.   Lantern Pinions:

This is a name given to an assembly consisting of two small hubs, mounted on an arbor, each provided with holes to receive pins, forming a cage or "lantern" which performs the same function as a solid pinion. Teeth of the mating wheel engage with the individual pins, one by one. In many lantern pinions, the pins are securely fixed in their holes. Others are designed so that the pins are free to rotate in their holes and are called "roller pinions".

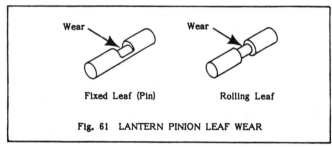

Fig. 61 LANTERN PINION LEAF WEAR

The pins in lantern pinions are susceptible to severe wear. Because of their relatively small diameter, this wear can result in bending or breakage. Bending may become severe enough to cause failure to engage with the teeth of the mating wheel. When a pin breaks, it may fall completely free of the pinion, or may become wedged against a tooth of the mating wheel. This condition may not be readily visible without disassembly. It can be detected, however, by using the tips of the fingers to slightly move the wheels in a train, one at a time. There

should be enough slack in the mating of wheels and pinions to permit a little motion. If there is none, it is an indication that a broken pinion pin may have caused that wheel to lock.

2.   Escape Wheel Teeth:

The escape wheel is the last wheel of the time train. It is usually near the top of the movement.

The function of the escape wheel is totally different from that of other wheels. Rather than transmitting power to another wheel, the escape wheel does so when one of its teeth impacts a pallet on the anchor and applies force, through the crutch, to give a slight push to the pendulum.

The shape of escape wheel teeth is very different from that of other wheels. They are thinner, more widely spaced and usually come to a sharp point. Each time the pendulum swings, one of the pallets of the anchor drops between two teeth of the escape wheel so that it is impacted by a tooth, stopping the movement of the wheel

While most escape wheel teeth are essentially triangular in shape, there are exceptions, notably in pin pallet (Brocot) escapements, where the pallets are round pins with flat backs, mounted perpendicular to the face of the anchor. In this case, the teeth are essentially rectangular, with a semi-circular cutout on the back side to form the tip.

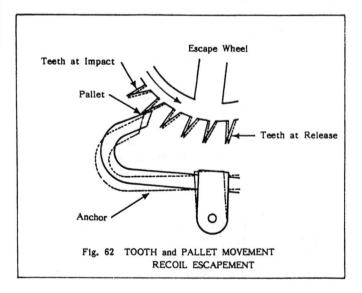

Fig. 62 TOOTH and PALLET MOVEMENT
RECOIL ESCAPEMENT

a. Recoil Escapement:

After a pallet is impacted by a tooth, the momentum of the pendulum exerts itself, momentarily, and causes the pallet to force the tooth backward very slightly, until the force imparted by the tooth pushes it in the opposite direction. It is this recoiling motion that gives this type of escapement its name.

As this is happening, the opposite pallet on the crutch is moving between another pair of teeth, where the same process occurs again. This alternating stop, power, release cycle is repeated continuously as long as the movement is running.

(1) Escape Wheel Tooth Damage:

(a)   Impact Damage:

The relationship between the tips of each of the escape wheel teeth and the pallets must be quite uniform for proper operation of the movement. Because the teeth are so thin, they are subject to damage. If wear causes the anchor arbor to become loose, or pallets and escape wheel teeth are badly worn, so that the pallets do not engage the teeth properly, the escape wheel can suddenly begin to spin rapidly. If, as the escape wheel is spinning, a pallet is lowered, the ends of the teeth will be bent as they strike it. When this happens, proper engagement is impossible and the clock will not run.

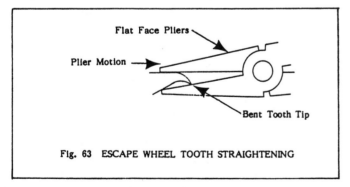

Fig. 63 ESCAPE WHEEL TOOTH STRAIGHTENING

(b)   Correcting Bent Escape Wheel Teeth:

In many cases, bent escape wheel teeth can be straightened, without disassembly of the movement. These teeth normally have two straight faces, forming a sharp

point. When they are damaged by bending, the tips will be curled slightly and the task is to straighten them, without breaking them off.

To perform this work, you must have a small pair of long nose pliers, with FLAT FACES on the inside of the jaws Those with serrated faces must not be used, because they will damage the surfaces of the teeth.

Select a tooth and, with the tip of the pliers just below the bent tip of the tooth, position one face of the pliers on the flat face of the tooth. Gently apply pressure, while pulling the pliers outward from the tooth, in a direction parallel to the flat face of the tooth. If done carefully, this will restore the point to its original shape.

Repeat this procedure one tooth at a time, until all teeth have been straightened.

If the anchor of your movement can be removed simply, as with those from American weight driven shelf clocks, remove it. While causing the escape wheel to spin at a fairly rapid rate by turning one of the large wheels in its train, hold a small piece of very fine (# 600) wet-or-dry abrasive paper, so that it just touches the tips of the escape wheel teeth. If the sound made as the sand paper touches the teeth is irregular, some of the teeth are longer than others. Do this a little at a time, visually checking the teeth frequently, until you hear a steady whir. This indicates the teeth are of uniform length.

In many cases, disassembly of the movement will be necessary in order to disengage the anchor and perform this operation. This will be covered in Chapter 11.

(2) Worn Escape Wheel Teeth:

Especially in recoil escapements, significant friction occurs between the teeth of the escape wheel and the pallets. After impacting the pallet, the tooth is forced backward slightly, with a rubbing action between its tip and the face of the

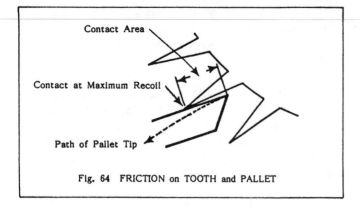

Fig. 64 FRICTION on TOOTH and PALLET

pallet This results in wear of both escape wheel tooth points and pallet faces. Most severe wear is usually evidenced by grooves worn in the steel pallets. The tip of the pallet is beveled, so that, as wear occurs, the bottom of the groove developed by wear effectively shortens the pallet.

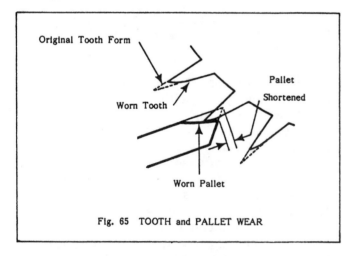

Fig. 65 TOOTH and PALLET WEAR

The combination of wear of both escape wheel teeth and pallets eventually reaches a point where, for a brief instant in the middle of the swing of the pendulum, no tooth is engaged by a pallet. This permits the escape wheel to run free. When the swing of the pendulum later causes a pallet to intercept the arc of the very thin spinning tooth tips, they are likely to be bent.

b. Recoil Escapement Anchor Variations:

The pallets of all recoil escapement anchors are flat and at an angle to each other such that each presents the same angle to the escape wheel tooth on impact. There is considerable variety in their shape, construction and method of

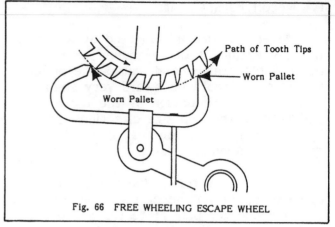

Fig. 66 FREE WHEELING ESCAPE WHEEL

mounting. In the illustrations that follow, note that, regardless of other differences, the faces of the pallet portions are identical. They are flat and essentially at the same angle to each other.

Another common characteristic lies in the fact that the tips are sharply angled. This is to allow them to drop between the teeth, without interference.

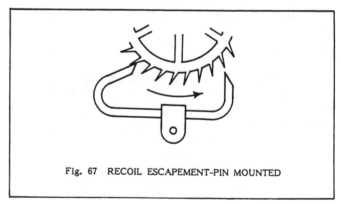

Fig. 67 RECOIL ESCAPEMENT-PIN MOUNTED

(1) Pin Mounted Recoil Escapement Anchor:

The anchor is made of a strip of steel, bent at each end to form a pallet. A tiny sheet brass saddle is riveted to the anchor and has holes to fit over a pivot pin. A wire crutch is also riveted to the anchor. The assembly fits outside the front plate of the movement and is usually positioned so that it is under, or to the side of, the escape wheel. This type is very common on older American clock movements.

(2) Arbor Mounted Recoil Escapement Anchor:

The anchor is made of strip steel,

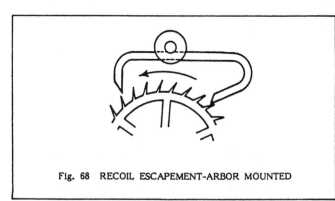

Fig. 68 RECOIL ESCAPEMENT-ARBOR MOUNTED

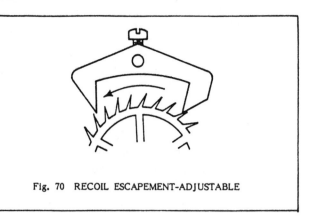

Fig. 70 RECOIL ESCAPEMENT-ADJUSTABLE

bent to form a pallet at each end. The arbor on which it fits is provided with a slot, into which the anchor is inserted and the edges of the slot crimped over to secure it. The arbor is positioned between the plates of the movement and the anchor is positioned over the escape wheel. This type is commonly found on American clocks made in the late 19th and early 20th centuries.

(4) Recoil Escapement- Adjustable

This form is similar to the fixed type, in that it is a heavy steel stamping or forging, but the fit between the anchor and the arbor is slightly looser, so that it may be moved. When the proper anchor position is achieved, a screw at the top is tightened to secure it to the arbor.

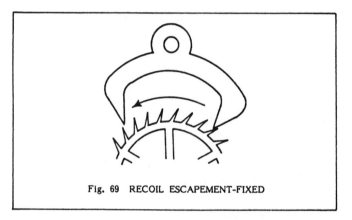

Fig. 69 RECOIL ESCAPEMENT-FIXED

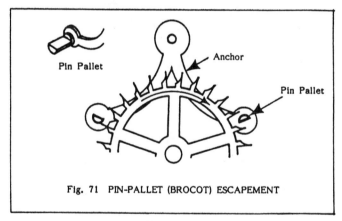

Fig. 71 PIN-PALLET (BROCOT) ESCAPEMENT

(3) Fixed Recoil Escapement Anchor:

Essentially similar in function to what we term the arbor mounted anchor, this type differs in that the anchor is made from a heavier steel stamping or forging. It is attached to its arbor by pressing the hole in its top over the arbor. The fit between the hole and the arbor is such that there is sufficient friction to secure one to the other.

In some cases, anchors that look much like this fixed one, may be subject to movement, with extreme care.

They fit very tightly on the arbor, but are not secured to it. They usually have a brass bushing, with flanges on each side. If adjustment is necessary, use only minimal pressure. Don't force it.

c. Pin Pallet (Brocot) Escapement:

This escapement is found on many fine clock movements of both American and European manufacture. It almost always is located in front of the dial and is sometimes referred to as a "Visible" escapement. It differs substantially from most other types in that its pallets are literally small pins, made of hardened steel, or semi-precious stone. These pins are mounted into an an anchor which is sometimes elaborately decorated.

The exposed end of the pins, which forms the pallet, has about half its cross section cut away to form a flat face. The escape wheel teeth impact the rounded side of the pins. Because of the thickness of the pins, the Brocot escape wheel teeth have very thin tips, formed by removing a circular section from the back side of an otherwise rectangular body. This is necessary so that the pallet pins can drop between the teeth.

The anchor is frequently attached to its arbor by a friction fit and can sometimes be adjusted to obtain proper beat. This must be done very carefully, however, to avoid serious damage.

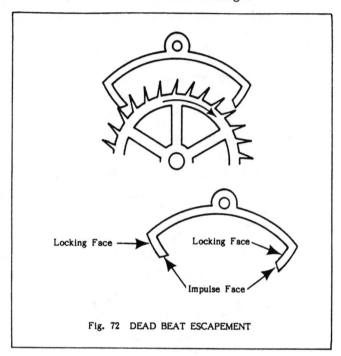

Fig. 72 DEAD BEAT ESCAPEMENT

d. Dead Beat (Graham) Escapement:

The principle difference between dead beat and other escapements is that the pallets are curved on a radius from the center of the anchor arbor. Thus, when a tooth comes into contact with a pallet, even though the pallet may continue to move, it exerts no force on the tooth. The tooth is simply locked and there is no recoil.

When the pendulum starts to swing in the opposite direction, moving the end of the pallet toward the tip of the tooth, the tooth comes into contact with the impulse face of the pallet and force is exerted to push the pallet upward. This provides the necessary impulse to the pendulum.

There being no recoil, or backward motion of the escape wheel, which remains

motionless for an instant after each locking, the name "dead beat" derives.

G.   PALLETS:

The pallets are the two surfaces at either end of the anchor which interrupt the movement of the escape wheel and transmit power to the pendulum. In recoil and dead-beat escapements they are usually a part of a single piece of steel bent or shaped to form a roughly "U" shaped element called the anchor. The anchor may have the crutch directly attached to it, or may be mounted on an arbor that also carries the crutch. In either case, the crutch then imparts power to the pendulum.

In some cases, however, the pallets may be separate pieces, attached to the anchor, as in a pin-pallet (Brocot) escapement.

It is not uncommon to find grooves worn in the pallets of old clocks. Curiously, even though the escape wheel teeth which impact the surface of the pallets are made of brass, wear is usually evidenced most severely on the pallets. This is accounted for by the fact that the pallets are impacted by every tooth on the wheel during each revolution, while individual teeth impact the pallets only twice each revolution.

Unless pallet wear is far advanced, it will probably not seriously affect operation of most movements. In extreme cases, correction or replacement will be necessary. We will discuss this in Chapter 12.

1. Pallet Engagement:

As we have indicated, the degree of engagement between the pallets on the anchor and the teeth of the escape wheel is important. Engagement is determined by the relative distance between the escape wheel arbor and the arbor on which the anchor is mounted. It must be deep enough so that one pallet is in position to stop a tooth the instant the other pallet releases the tooth it has been engaging.

If there is significant wear of pallets and/or teeth or if the arbors are too far apart, due to pivot hole wear, the pallets will not enter deeply enough between the teeth to lock and release as they should. As a result, there is a period near the center of the swing of the pendulum when neither pallet engages a tooth, so the wheel is free to spin, which it will do. As the pendulum continues its motion, one pallet starts to enter, but is likely to bend tooth tips before it drops deep enough to stop the wheel.

If the arbors are too close together, one pallet will be stopped at the bottom of the space between two teeth on the escape wheel before the other pallet can release the tooth bearing against it, so that the wheel cannot turn.

Adjusting Pallet Engagement:

CAUTION: It must be assumed that the movement you are concerned with once operated satisfactorily and that the engagement of the pallets with the escape wheel teeth was originally correct. Any adjustment made now is only to compensate for wear or after repair of damaged teeth, so only slight change in the position of the arbor is needed.

It is important to emphasize that any means of adjustment provided was probably intended to be used only when the movement was originally assembled. In some cases, escape arbor pivot supports are retained by screws which can be loosened, but in most cases, adjustment now will mean bending or otherwise forcing a change of position of the pivot hole or pin.

Note that adjustment of pallet engagement will not satisfactorily correct for pivot hole wear.

Before attempting any adjustment, be certain that such adjustment is necessary and will overcome the problem. Be aware that only very slight change is required. Don't overdo it.

After each adjustment, check the movement of the escape wheel as the pallets are rocked back and forth.

Provision is usually made, in the design of the movement, to permit some adjustment, at the time of manufacture, of

the location of the position of the arbor on which the anchor is mounted. This takes many forms:

(1) Separate Anchor Pivot Supports:

Perhaps the most common form of anchor pivot support is the attachment of separate brass brackets to the movement plate. This part may carry a pin for the pivot holes of an anchor mounted in front of the plate, or may have a hole for the pivot of an anchor arbor carried between the plates. It is usually secured by a single rivet to the movement plate. Occasionally screws may be used, instead of rivets. Some movements have a bracket only on one plate, others have one on each plate.

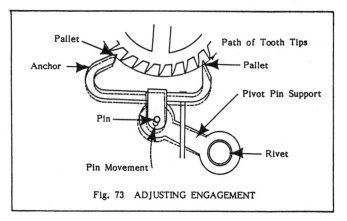

Fig. 73 ADJUSTING ENGAGEMENT

(a) Adjustment of Pivot Supports:

If the support bracket is attached by a screw, barely loosen it, using a screw driver, then rotate it very slightly toward the escape wheel.

If the support piece is secured by a rivet, remembering that this piece is made of soft brass, use the nose of the jaws of your flat nose pliers to grip it, making sure the pliers are in good contact with the enlarged portion at each end of the support piece. Using only necessary force, turn in the direction that will bring the anchor arbor closer to the escape wheel arbor. Only very slight movement is necessary, as shown in Fig.71.

(2) Bridge Pivot Support:

Particularly on old English movements, when the anchor is inside the plates, but the crutch must be outside, one end of the

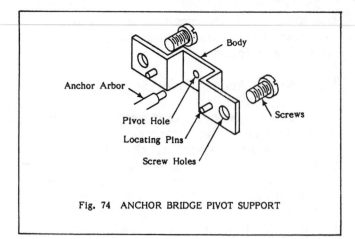

Fig. 74 ANCHOR BRIDGE PIVOT SUPPORT

arbor may be supported in a pivot hole in a bridge attached to the outside of the backplate. Fortunately, because of their substantial construction, wear is minimized and adjustment is seldom necessary.

(a) Adjustment:

In most cases, the maker has installed one or more steel locating pins in the feet of the bridge, which fit closely into holes in the plate, fixing the position of the pivot hole. The bridge is secured with screws. Little or no adjustment is possible and, in the rare instance when correction is necessary, it will probably require rebushing or other repairs which should only be done by a skilled clock maker.

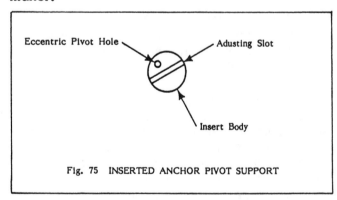

Fig. 75 INSERTED ANCHOR PIVOT SUPPORT

(3) Inserted Anchor Pivot Support:

Especially in French clock movements, a round piece of brass in which the pivot hole is located off-center is tightly inserted into a hole in the plate. A slot, like that in a screw head, is cut into the face of this round piece, so that it may be rotated slightly to change the pivot hole position.

(a) Adjustment:

Using a sharp screwdriver that closely fits both the width and length of the slot, slightly rotate the insert in the direction that will bring the pivot hole closer to the escape wheel arbor. It will be necessary to apply substantial pressure, so that the screw driver does not slip out of the slot. If it does, the edges of the slot can be so damaged as to make it unusable and the plate will be seriously marred. BE CAREFUL.

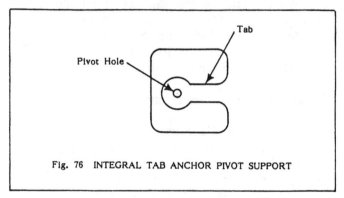

Fig. 76 INTEGRAL TAB ANCHOR PIVOT SUPPORT

(4) Integral Tab Pivot Support:

This form consists of a tab, containing the pivot hole, which is part of the movement plate, formed when a shaped hole was punched in the plate.

(a) Adjustment:

Place the edge of the tip of a small screw driver inside the hole, on the side opposite the direction in which you want to move the end with the pivot hole, with the face of the screw driver bearing against the edge of the tab. Apply slight pressure to the end of the tab by twisting the screw driver. Do this a little at a time, checking engagement of the pallets each time. DON'T OVERDO IT. If you do bend it too far, so that the pallets engage too deeply, place the screw driver in the opposite side of the hole and bend the tab back. If this is done very often, the tab may break off.

H. PIVOTS and PIVOT HOLES:

Pivots are the small ends of arbors which act as the male part of the bearing permitting rotation or oscillation of the arbor. In better movements, they are

74

hardened and highly polished to provide a surface generating a minimum of friction. In movements of lesser quality, the surface is not as smooth and more friction is generated.

Pivots are made of steel and nearly always run in a pivot hole of brass. While pivots are subject to wear, the extent of wear is usually much less than that of the holes in which they run. Sometimes, however, abrasive particles of dust become embedded in the brass of the pivot hole and cause grooves to be cut in the pivot.

Pivot holes are the holes in the plates of the movement into which the pivots fit and which form a bearing. Accurate location of these holes is essential to insure proper engagement of wheel and pinion teeth, as well as pallet and escape wheel teeth. While the pinions are made of steel and the plates of brass to minimize friction, wear does occur, particularly in the holes.

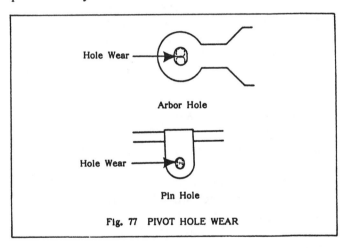

Fig. 77 PIVOT HOLE WEAR

As wear progresses, it is evidenced by enlargement of the pivot holes, which tend to become oval in shape. When the extent of wear is excessive, engagement of the teeth is affected and can result in stoppage of the movement.

1. Pivot Hole Wear:

a. Causes of Wear:

Pivot hole wear results from friction between the surfaces of the pivot and the hole in which it turns. If both surfaces are highly polished, fit closely, but

freely, are long enough to provide good bearing area and are properly lubricated, wear is greatly minimized.

If pivots do not fit closely in their holes, it is likely that small particles of abrasive dust will find their way between the two surfaces and increase the rate of wear. Oil that effectively adheres to the surface of the pivot and its hole tends to exclude such dust, as it provides lubrication. Too much oil, however, tends to attract dust which inevitably finds its way into the bearing area and aggravates wear.

As wear progresses, the space between the bearing surfaces increases, more dust is admitted and wear is accelerated.

b. Effects of Wear:

Some types of movements can tolerate more wear of pivots and pivot holes than others. In general, it may be said that movements with coarsely formed wheel teeth, large pivots and relatively thin plates, are more likely to evidence substantial pivot hole wear. Such movements frequently continue to operate even when there is significant wear.

Movements with thick plates, smaller highly polished pivots and finely cut teeth generally evidence much less wear at the pivot holes. However, because the teeth are designed to fit much more precisely, significant pivot hole wear can prevent teeth from meshing properly, resulting in stoppage.

Greatest wear normally occurs in pivot holes of pivots that make the greatest number of revolutions in a given period of time. The arbors of the escape wheel in the time side train and the fly governor in the strike side, the last in each train, make a great many more rotations than any others and are, therefore, likely to cause much more wear of their pivot holes.

When springs or weights exert more force than necessary, wear of other pivots and pivot holes can also result from a combination of friction and excessive pressure. Because of improper design, some movements were overpowered at the time of manufacture, frequently to compensate for friction resulting from crude construction details.

In many cases, heavier springs or weights have been added by lazy or inexperienced repairers in an often successful effort to make a clock run, without proper repair. Overpowering inevitably results in severe wear and can cause significant damage.

Pivot holes of the anchor arbor, or pivot bracket, which oscillate once for each tooth on the escape wheel, very frequently exhibit significant wear.

2. Elongated Pivot Holes:

Take a moment to examine each pivot hole on the front and back plates of your movement. In many cases, the pivot will extend beyond the face of the hole so that you can touch it with the tip of a finger. Gently try moving it around in the hole to see if it is loose. There will be some movement, but if it appears to be excessive, look closely at the hole. If the hole is elongated, or oval, it may very well be a source of trouble.

Some movements seem to run quite satisfactorily, even when one or more pivot holes appear to be significantly enlarged. Check the engagement of the wheel and pinion teeth on the affected arbors. If they seem to function reasonably smoothly, they will probably operate satisfactorily for a reasonable period of time. If there is interference, the problem must be corrected.

a. Correcting Worn Pivot Holes:

Proper correction requires not only that the size of the hole be reduced to fit the pivot, but that it be centered in order to assure proper engagement of the teeth.

As you examine your movement, you may find evidence of little dimples, or punch marks, or an indented ring around some pivot holes. This was done in an effort to squeeze the metal around the hole to make it smaller. Known as "hole closing",

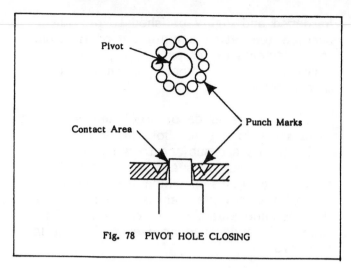

Fig. 78 PIVOT HOLE CLOSING

this only alters the extreme outer edge of the hole, leaving a very thin bearing area which will wear rapidly. This is a very poor practice, which should never be repeated.

(1) Rebushing Pivot Holes:

The only satisfactory way of repairing a severely worn pivot hole is to rebush it. The original center of the hole must be located and a larger hole made. Into this hole a tightly fitting brass bushing is pressed. The bushing has a hole in its center which must be of the right size to provide a proper fit for the pivot.

Clock makers are prepared to do your rebushing, or you may find someone in your NAWCC Chapter who can do it for you.

Bear in mind that disassembly and reassembly of the movement is generally necessary. This is time consuming and, if you don't do it yourself, must be paid for.

In Chapter 12, we will discuss a good method of rebushing requiring a minimum of additional investment in tools which you may want to try, after you have disassembled your movement.

3. Pivot Polishing:

Whenever elongated pivot holes are encountered and it is necessary to rebush them, the mating pivot should be carefully inspected for roughness, perhaps caused by rust, and grooves which have resulted from abrasion. We will discuss correction of these conditions in Chapter 12.

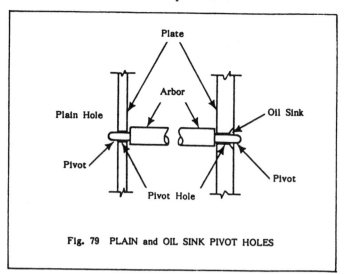

Fig. 79 PLAIN and OIL SINK PIVOT HOLES

4. Pivot Hole Lubrication:
a. Oil Sinks:

As with any moving parts, proper lubrication of pivots is essential to minimize friction and wear. Because of the very small size of clock pivots, use of special oils that will adhere to the working surfaces for long periods of time, without migrating away from the bearing, are essential.

Better quality movements nearly always are provided with what are known as "oil sinks" at each pivot hole. These sinks are tapered depressions in the plate, around the pivot hole and are designed to help retain a minute quantity of oil at the pivot.

Many movements have no sinks at all, and some have them only on the face of plates that are readily visible, with none on the plate that is inaccessible without removal from the case.

b. Lubricating Pivots:

(1) Clock Oil:

Use ONLY oil specially formulated for use on clock movements, which is available from clock parts suppliers.

(2) Applying Oil:

DO NOT OVER OIL! Only the tiniest bit, much less than a drop, of oil should be applied to each pivot.

When oil sinks are provided, slightly more oil may be used, but must never go beyond the edges of the sink.

Place the movement with the back plate on the work surface, so that the front plate is up and horizontal.

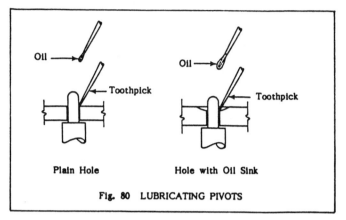

Fig. 80 LUBRICATING PIVOTS

Dip only the very point of a toothpick in the clock oil. Touch it to the lip of the oil bottle so that only a tiny amount is left on the tooth pick. Carefully touch it to the side of the pivot, just above the hole, then move it down to the hole. No oil should flow onto the plate. If this does happen, quickly wipe it with a soft cloth or lintless tissue. Repeat for each pivot. Then turn the movement over and oil the pivots on the back plate in the same way.

I. LUBRICATION of PALLETS, CRUTCH and LEVER ARBORS:

These are the only points, other than the pivots, that should be lubricated.

A. PALLETS:

Using a toothpick again, dip the point in clock oil and wipe it on the edge of the bottle as you withdraw it, so that very little oil is left on it. Touch it to the faces of the pallets, where they come in contact with the escape wheel teeth.

B. CRUTCH:

With the same procedure as described for pallets, touch the toothpick to the inside faces of the crutch loop, in the case of wire pendulum rods, or to the sides of the slot when there is a crutch pin, or to the sides of flat crutches, where they come in contact with the pendulum connection.

C. LEVER ARBOR PIVOTS:

Treat lever arbor pivots in the same way as for the pivots of the time and strike trains.

Chapter 9

STRIKING MOVEMENTS

Strictly speaking, only timekeeping devices capable of announcing the hours audibly are "clocks". Those which do not do so are called "time pieces". Common abbreviations are T/O (Time Only) for time pieces and T/S (Time and Strike) for true clocks.

Up to this point, we have been concerned with problems common to both the time and strike trains of the clock movement. While we have mentioned the strike train, we have not looked at it closely. It performs functions totally different from those of the time train, while having a similar source of power and a series of wheels and pinions to transmit power. Here we will briefly examine the features of the strike train that differ from, yet interact with the time train.

The basic function of a strike system is to provide an audible announcement of the time. This requires that it be activated by the timekeeping mechanism, so that the announcement is made when the hands show the time struck. Striking must be synchronized with the visual time indicating function of the clock, i.e., striking the hour when the minute hand reaches the 12 o'clock position.

I. STRIKE SYSTEMS:

A. PASSING STRIKE:

In the simplest form of striking clock, a gong, bell or other sound generating device is struck only once as the minute hand arrives at 12 on the dial. This is accomplished by a pin or similar projection on the minute arbor of the time train, which raises the end of a pivoted lever with a hammer on its opposite end. As the pin continues its travel past the end of the lift lever, it will drop and the hammer will strike the gong or bell once when the minute hand reaches 12. This is called a "passing strike". It does not count the hours.

Since a weight driven time piece with no strike, or with passing strike, requires only one weight, it is commonly referred to as a "one weight" clock.

B. COUNTING STRIKE SYSTEMS:

More complex forms of strike systems count the number of hours indicated by the hands. They nearly always require a separate power source and train of wheels, capable of storing enough power to lift and drop the strike hammer 156 times in each 24 hour period. The strike train must be activated by the time train, at the hour.

Having two trains, weight driven clocks that strike must have two weights and are referred to as "two weight" clocks.

C. COUNT WHEEL STRIKE:

Perhaps the most common strike system is called "count wheel strike". This name derives from the fact that a special wheel is added, usually on the same arbor as the first or "great" wheel of the strike train. This wheel has 78 teeth, the spaces between 12 of which are much deeper than the rest. When the strike train is activated, the end of a lever is dropped into the space between teeth, progressively as the wheel rotates. The strike train will continue to run as long as the lever rests in the shallow spaces between teeth. When the lever falls into a deep space, it causes the strike train to stop. The number of shallow spaces between deep ones determines the number of times the strike lever hits the gong or bell.

In addition to striking the hours, count wheel strike systems may also strike a single note at the half-hour. A second lifting device is attached to the minute arbor of the time train, diametrically opposite to the one activating the hour strike, to activate the train at the half hour. 12 deep spaces are added to the count wheel, each located next to the ones ending the hour strike sequences. At the half hour, the count lever drops into a deep space after one strike and the strike train stops.

Sometimes the half-hour is struck on a separate bell or gong. When this is the case, a separate strike lever is activated by a second pin attached to the minute arbor of the time train, diametrically opposite and offset from the one activating the hour strike. This pin lifts and drops the second lever at the half hour.

In most cases, all of the elements, levers, cams, wheels, etc., of a count wheel strike system are located between the plates of the movement. The count wheel of wood movements is usually located outside the plate.

The operation of Count Wheel Strike Systems will be explained in further detail later in this Chapter.

D. RACK and SNAIL STRIKE SYSTEM:

This system gets its name from a toothed semi-circular element called the "rack", which is advanced one tooth for each hour struck, and a stepped cam, resembling the shell of a snail, which determines the number of teeth to be activated in the rack.

Usually, many of the elements of a rack and snail system are located outside the plates of the movement.

The operation of Rack and Snail strike systems will be explained in more detail later in this Chapter.

E. COUNT WHISTLE (Cuckoo):

In addition to striking the hour on a gong, the familiar cuckoo clock also announces the hours by successive activation of two whistles, each sounding a different note, to simulate the familiar call of the cuckoo bird.

This is achieved by wooden whistles, on top of which are relatively large bellows. The top of the bellows is lifted and dropped, to provide a puff of air for the whistle, by a lever activated by the strike train of the movement. Usually, there is also a strike lever which simultaneously strikes a wire gong mounted on the back cover of the case.

F. QUARTER HOUR STRIKE:

Still more sophisticated systems, in addition to striking the hour, include a means of announcing the quarter hours. The simplest of these, strike once at 15 minutes past the hour, twice at 30 minutes, three times at 45 minutes and four times preceding the hour strike.

Such systems are usually operated by a third train in the movement. Quarter hour striking is usually sounded on a bell or gong which differs in tone from the one sounding the hours.

Some cuckoo clocks announce quarter hours in similar manner, except that, in addition to striking a second gong, a third whistle is activated to produce three short notes for each of the one to four required for successive quarters, to simulate the call of the bob-white quail. This requires a third train and a third weight.

Other clocks strike two short notes for each element of the quarters, on the same bell or gong that announces the hours.

Since three trains are involved, weight driven clocks with quarter hour strike have three weights. They are

80

commonly referred to as "three weight" clocks.

G. QUARTER HOUR MELODY CHIMING:

Chiming clocks play a part of a melody at the quarter hour, adding successive parts at succeeding quarters, until the full melody is played at the hour. These chime systems, like quarter strike movements, require a third source of power and train of wheels and pinions.

Chiming clocks have a series of tuned rods, tubes, or bells, each with a separate hammer. There is a separate hammer for each. Set into a drum driven by the chime train are a series of pins in line with the tail of each hammer lever. The spacing between each pin in a series, determines when the hammer it activates will be lifted and dropped to sound its particular note. In this way, each hammer is programmed to sound its note in proper sequence to become part of the melody.

Some chime movements are capable of playing several different tunes. This is accomplished by having two or more series of pins on the drum for each hammer. By moving the drum slightly from side to side on its shaft, a different set of pins, spaced to produce another tune, is brought into alignment with the tails of the strike hammers.

Full melodies used on chiming clocks consist of four parts, each one a recognizable tune. With the common Westminster melody, the first part, played at 15 minutes past the hour, consists of 4 notes. The second, played at the half hour, has 8 notes. The third, played at 45 minutes past the hour, has 12 notes. The full melody, played just before the clock strikes the hour, has 16 notes.

The chime drum, then, must be started, run for the required period at each quarter hour and then be stopped. It must be properly synchronized, not only with the time, but with the exact part of the melody specified.

H. MUSICAL CLOCKS:

The simplest form of musical clock is found in those made in the Black Forest of Germany, notably of the Cuckoo variety. They are equipped with a small music box movement that plays a readily recognized melody at the hour, and sometimes at the half hour. The sound is made by tiny fixed steel prongs of varying length that are pricked by properly placed steel pins on a small rotating drum, causing the prongs to vibrate.

While they are so rare that you are unlikely to have one in your collection, it may be of interest to know that there are clocks that play longer and more complex musical selections on bells, rods, or tubes and sometimes on organ pipes. In principle, they operate like chiming clocks, but have much larger drums, with space for many more pins. Several tunes may be incorporated on a single drum and may be played in sequence, or a particular melody selected. The entire melody is usually played on the hour.

I. SUMMARY:

Because some of these systems are mechanically complex, beyond the scope of this book, we will not go into them further here. You may want to consult some of the more advanced repair books listed in the bibliography, some of which contain further information about strike systems.

If you have a chiming clock of recent manufacture, much helpful information may be found in the owner's manual provided by the manufacturer.

II. COMPARISON of STRIKE TRAINS and TIME TRAINS:

A. COMMON FEATURES:

1. Power Source:

All strike trains must have a source of power, be it a spring or a weight. Ideally, the power supply should provide only enough power to efficiently operate the movement. Too little power will not allow proper operation, while too much will cause excessive wear. In general, the power supply for the strike train is likely to be identical to that used for the time train. In well designed movements, this is not always the case, since the design of the movement may call for more power to effectively operate one train than the other. To avoid excessive wear resulting from over powering, springs in the same movement may have very different characteristics, or weights may be of different weight for each train.

2. Wheels and Pinions:

In most cases, the wheels and pinions of the strike train will very closely resemble those of the time train, although they are likely to vary considerably in the number of teeth on each wheel or pinion.

3. Arbors and Pivots:

Usually arbors and pivots of the strike train will be very similar, if not identical to those of the time side.

B. UNIQUE FEATURES OF STRIKE TRAINS:

1. Occasional Operation:

Unlike the time train, whose motion is interrupted only for an instant by the escape system, the strike train is completely at rest most of the time. It operates only when activated to sound the strike and then only for the time it takes to perform that function.

2. Work Done:

While the time train uses its power, a little at a time, to impulse the pendulum and turn the hands, the strike train must exert greater bursts of power to lift the hammer and control the number of times this is done, all in a very short period of time. It must accelerate from a standing stop, to full speed, very quickly. In addition, having no escape wheel, it must have a governor to control the speed at which power is released.

3. Lever Controls:

Control of strike train functions, starting, striking and stopping, is done by levers mounted on separate arbors,

activated by pins and/or cams mounted on some of the train arbors.

4. Count Systems:

The strike train must include a means of determining the number of times the hammer will strike each time it is activated. We have already mentioned the two most common systems for accomplishing this, Count Wheel and Rack and Snail. Now, we will take a close look at their elements and how they operate.

III. FUNDAMENTAL STRIKE SYSTEM MECHANICS:

The basic nature of this book will limit us to consideration of only the most common strike systems found on clocks, touching on minor difficulties with them, which can be corrected simply.

A. BASIC FUNCTIONS:

Certain actions are necessary with any strike system:

1. Locking:

As soon as the proper number of strikes has been completed, the strike train must be stopped and kept from moving until it is time for the next hour or half hour to be struck. This is usually done by a lever which moves to interfere with a pin projecting from the side of the last wheel in the strike train, or into a slot in a cam attached to a wheel, thus stopping all motion of the train.

2. Warning (Cocking):

A few minutes before the hour is to be struck, the locking lever is lifted up by a cam or pin on the minute hand arbor which, simultaneously, lifts another lever up so that a tab on its end interferes with a pin on another wheel. This "cocks" the movement, so that it is ready to be instantaneously activated at the hour.

3. Releasing:

The moment the minute hand reaches the hour and the warning lever drops clear of the lifting pin, the train is released to start striking. At the end of the strike cycle, the locking lever again moves and stops the strike train.

4. Counting:

Since the purpose of the strike train is to indicate the hour audibly, there must be a mechanism for determining the number of times the gong or bell will be struck.

5. Striking:

A means must be provided to lift and release a hammer so that it strikes a bell, gong, or rod.

6. Speed Control (Fly Governor):

When in operation, the strike train runs continuously and a means for controlling its speed at a uniform rate must be provided. In nearly all strike systems, this is achieved by use of a thin flat metal piece, called a "fly" (fan) mounted on the last arbor of the strike train.

Some clocks, notably Japanese and Korean clocks of recent manufacture, use centrifugal elements inside a fixed drum to control speed by a braking action.

Since the strike train must accelerate from a dead stop to controlled speed in a short time, the fly is usually held in place on the arbor by a fine spring that fits into a groove in the arbor. This spring acts as a clutch and allows the arbor to accelerate rapidly, while air resistance on the fly causes it to pick up speed a little more slowly. This clutch action is essential to proper operation of trains in which it is used.

If the fly spring does not fit snugly against its arbor, power will not be transmitted to the fly blade and there will be little or no braking action. In this case, the clock will strike more rapidly than it should. Bending the blades of the fly so as to increase tension on the spring sometimes helps. Only minimal pressure of the spring against the arbor is necessary and it should turn easily, but not be so free

that it spins rapidly when flipped with a finger tip.

B. COUNT WHEEL (Locking Plate) STRIKE SYSTEM:

Commonly called Count Wheel in the United States, In Europe this is known as the Locking Plate System.

This is probably the oldest and most widely used system of striking, no doubt because it is essentially simpler, with fewer parts and therefore less expensive to produce than the rack and snail system. It is by far the most common strike mechanism found on American movements.

Count wheel striking has a serious drawback; it operates without regard to the position of the hour hand and not infrequently gets out of synchronization so that an hour other than the one indicated by the hand is struck. In most cases, this lack of synchronization can be corrected by simply moving the hour hand to the hour being struck, then advancing the minute hand to set the proper time.

When resetting the time by moving the minute hand, it is essential to pause and allow the clock to strike at each hour. This is not usually necessary with rack and snail strike systems.

While details of the elements of a count wheel strike system will vary with movements of different design, the principles of operation will generally conform with the following description.

1. The Count Wheel:

This system derives its name from a large toothed wheel, sometimes called the locking plate, but more commonly known as the count wheel. Its function is to determine the number of times the gong is struck each time the strike train is activated. Each tooth on this wheel, or more accurately, each space between teeth, represents one strike. In twelve hours, a clock strikes a total of 78 times. If the count wheel is designed to strike only the hours, there will be 78 teeth, and spaces between teeth.

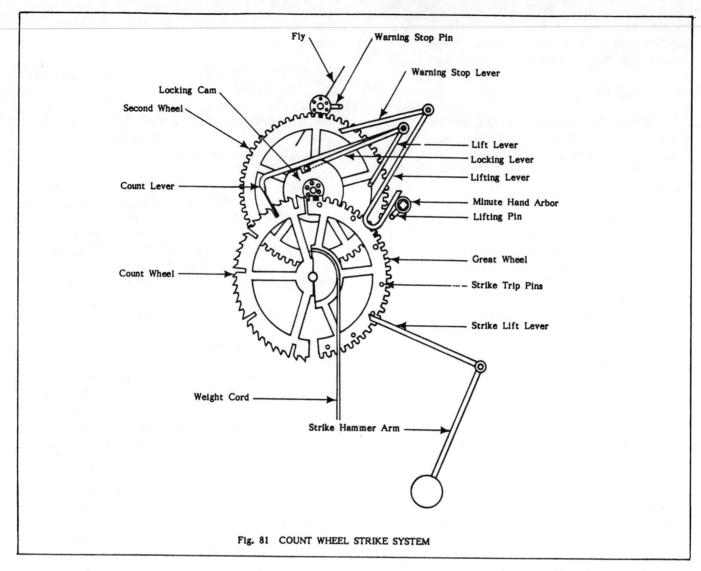

Fly

Warning Stop Pin

Warning Stop Lever

Locking Cam

Second Wheel

Lift Lever

Locking Lever

Lifting Lever

Count Lever

Minute Hand Arbor

Lifting Pin

Great Wheel

Count Wheel

Strike Trip Pins

Strike Lift Lever

Weight Cord

Strike Hammer Arm

Fig. 81 COUNT WHEEL STRIKE SYSTEM

If the clock also strikes the half hours, using the strike train, there must be an additional 12 spaces, for a total of 90. The train shown in Fig. 81 strikes only the hours. If it struck half hours, there would be another deep slot next to each of those shown.

In the weight driven movement shown, the count wheel is mounted on the same arbor as the first, or great wheel. In the figure, the count wheel appears at the left and the first wheel at the right.

To separate the hours, and provide a means of stopping the strike cycle at the end of a given sequence, deep slots are provided, instead of the usual shallow spaces between teeth. It is these deep slots that make the count wheel easily distinguishable from other wheels in the movement.

Looking at the top of the count wheel, you will find there are three teeth with shallow slots between them and a deep slot on each side. Then there are four teeth followed by a slot. This continues to the greatest space between slots, which has twelve teeth.

The count wheel makes one revolution in twelve hours, as it controls the striking of hours one through twelve.

2.   The Count Lever and Lock Lever:

Because these levers are attached to the same arbor and move together, we will discuss them together here.

a.   The Count Lever:

Mounted on an independent arbor, separate from the strike train, the count lever has a 90 degree bend at its outer end. The tip of the bent down portion is flattened and widened, so that it is much thinner than the space between teeth on the count wheel. This end is positioned so that the flattened end is centered on the rim of the count wheel and is radial to the center of that wheel.

The function of the count lever is to permit the strike train to run and strike the number of times set by the number of tooth spaces between the deep slots. It is lifted by a cam and lowered into the space between teeth with each strike.

When a deep slot is reached, the end of the lever falls into it and the lock lever drops into a slot in the cam to stop the train. Fig. 81 shows the count lever in a deep slot of the count wheel and the locking lever in the slot of the cam, locking the strike train.

b. The Lock Lever:

This is a shorter lever, mounted on the same arbor as the strike lever. It has a short bend at its end which, when the end of the count lever drops into a slot, moves into position to lock the strike train, usually by dropping into a slot in a cam mounted on another arbor, and engaging a face of the slot.

By gently lifting and lowering the count lever, you will be able to observe the motion and locking mechanism at the end of the lock lever.

c. Locking Cam:

Usually mounted on the second or third wheel of the strike train, is a locking cam. This may be a circular brass disc with a flat side and a rectangular slot, as shown in Fig. 81, or it may be snail shaped with a curved section ending at a flat face radial to the arbor.

The cam is shaped so that when the count lever is stopped at the bottom of the shallow spaces between the teeth on the count wheel, the end of the locking lever is held just below the top of the slot of the cam.

As the cam rotates, a rounded corner of the slot in the cam lifts the end of the locking lever, moving the end of the count lever up so that it clears the next tooth on the count wheel. This action is repeated until a deep slot in the count wheel is reached.

When the end of the count lever drops into a deep slot of the count wheel, the end of the lock lever drops deep into the slot of the rotating cam, effectively stopping it and the rest of the strike train.

3. The Warning Stop Lever and its Lifting Lever:

Since these levers are usually mounted on the same arbor we will discuss them together here.

a. The Warning Stop Lift Lever:

The function of this lift lever is to raise the end of the count lever out of a deep slot in the count wheel and the locking lever out of the lock position in the cam slot, at the same time moving the end of the warning lever into its locking position.

The lift lever is usually mounted on the same arbor with the warning lever. As it is raised by a lifting pin on the minute arbor, it comes in contact with the bent end of the locking lift lever to raise the count and locking levers.

Lifting and dropping is done by a pin or cam mounted on the minute (center) arbor of the time train. The lever is lifted once each hour, or every thirty minutes for half hour striking, in which case, there will be two cam points or pins on the minute arbor, one longer than the other.

When the pin or tip of the cam moves past the end of the lift lever, it drops, along with the warning lever, and the strike train is activated.

b. Warning Lever:

This lever is usually mounted on the same arbor as the lift lever. Its function is to temporarily stop the motion of the strike train by interfering with the rotation of the fly arbor, shortly after the lock lever and the count lever have been lifted out of their respective slots.

As the lock lever is released from the slot in the cam, the train will start in motion, but the warning lever has already come into position to interfere with the warning stop pin on the fly arbor to again stop the train, before striking can begin.

The lock lever will usually interfere with a pin on the last wheel or arbor of the train, but may sometimes act directly on the blade of the fly itself.

When the lifting lever drops off the lifting pin or cam, the warning lever drops clear of the warning stop pin and the strike train begins to operate. Striking starts and will continue until the count lever drops into a deep slot in the count wheel and the lock lever drops into the slot in the cam to again lock the strike train.

4. Strike Lift and Hammer Levers:

A third independent arbor carries the strike lift lever (sometimes called the hammer tail) and strike hammer arm. The strike hammer arm has a hammer at its end for striking the gong or bell. The strike lift lever is lifted by pins on one of the wheels, as in Fig. 81 or by the arms of a star wheel cam, then dropped to perform striking.

5. Lever Springs:

Many strike train levers depend only on gravity to drop them into position. Each of the lever arbors of the strike system may be provided with a light coil spring to augment gravity and insure rapid action. These springs are made of spring brass or steel coiled around the arbor, with one end hooked around the base of a lever and the other hooked over the edge of a movement plate.

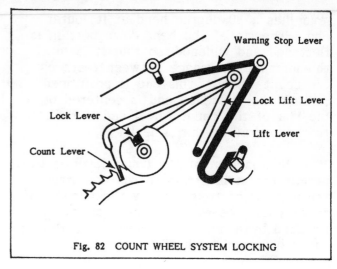

Fig. 82 COUNT WHEEL SYSTEM LOCKING

6. Count Wheel Striking Sequence:

a. Locking:

Except during the strike sequence, the strike train is locked as shown in Fig. 82. The end of the count lever is in a deep slot in the count wheel, allowing the lock lever to rest in the bottom of the slot in the cam, preventing it from turning and locking the train. The end of the lift lever rests on the minute arbor.

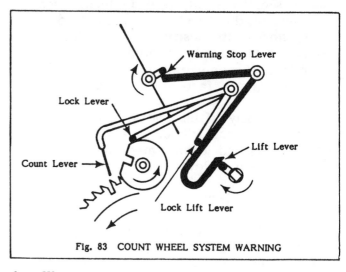

Fig. 83 COUNT WHEEL SYSTEM WARNING

b. Warning:

As the time train causes the minute arbor, carrying the minute hand, to rotate, the lifting pin mounted on it gradually raises the lift lever.

As the lift lever is raised, it comes in contact with the lock lift lever which begins to move upward. The count lever starts to rise out of the slot in the count wheel and the lock lever begins to move out of the slot in the cam. At the same time, the warning stop lever is raised.

As the minute hand approaches a few minutes before the hour, the lock lever clears the slot in the cam and the strike train is free to move. The cam begins to rotate, as does the fly. However, by this time the warning stop lever has moved into position to interfere with the warning stop pin on the fly arbor and the movement stops. A slight click can be heard at this time, as the stop pin strikes the warning stop lever.

Note that the lock lever now rests on the face of the cam, holding the tip of the count lever clear of the tips of the teeth of the count wheel.

The strike train is now cocked and ready to start striking. Why this phase is called "warning" is unclear.

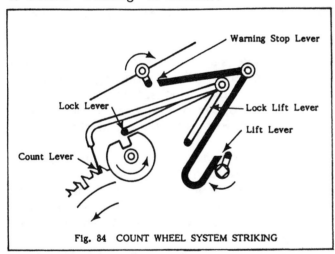

Fig. 84 COUNT WHEEL SYSTEM STRIKING

c. Running (Striking):

The moment the minute arbor brings the minute hand to the 12 o'clock position, the lift lever falls clear of the lift pin on the minute arbor and its tip comes to rest on the minute arbor.

As the lift lever drops, so does the warning stop lever, releasing the fly arbor and allowing the strike train to begin moving and striking to start.

When the locking cam and count wheel rotate, during striking, the count lever drops into one after another of the

shallow slots in the count wheel. The lock lever is thus prevented from dropping into the slot in the cam as it passes. As rotation continues, the flat face of the cam lifts the locking lever, causing the count lever to rise and ride on the face of the cam. This keeps the end of the count lever clear the tip of the next tooth on the count wheel.

The train is designed so that the strike hammer is lifted and dropped simultaneously with each stroke of the count lever.

This action is repeated until the proper number of the hour has been struck and the count lever drops into a deep slot in the count wheel, allowing the lock lever to enter the slot in the cam, locking the strike train, as shown in Fig. 82. It will remain locked until the next hour, when the cycle will be repeated.

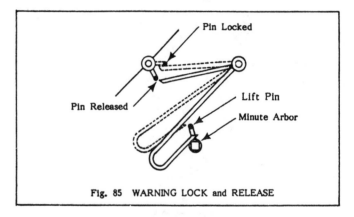

Fig. 85 WARNING LOCK and RELEASE

d. Warning Lock and Release:

The dotted lines of the levers in Fig. 85 show their positions when the lift lever is supported on the lift pin, just prior to release. The warning stop lever locks the pin on the fly arbor. As the minute arbor rotates, the end of the lift lever is released and falls onto the minute arbor.

Note that the arc of rotation of the tip of the lift lever brings it under the lift pin.

C. RACK and SNAIL SYSTEM:

This system provides a major advantage over others, in that the position of the hour hand is fixed with respect to the number of times the strike will sound when it is positioned at a given number on the dial.

For example, when the hour hand is properly synchronized and pointing to the three on the dial, when the minute hand reaches twelve, the clock will always strike three times. As the minute hand again approaches twelve, the hour hand will point to four and four strikes will be sounded, etc.

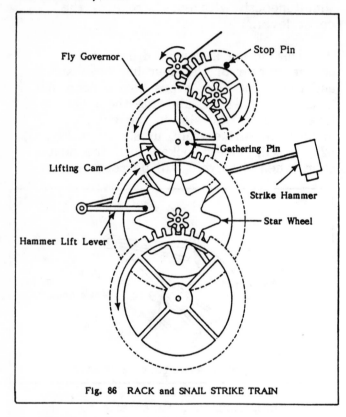

Fig. 86 RACK and SNAIL STRIKE TRAIN

1. Rack and Snail Strike Train:

This system is seldom found in 30 hour movements.

Typically, the strike train has four wheels and a fly governor. Power is applied at the first wheel, either by a weight or by a spring. A star wheel is commonly added, to provide the lifting and dropping action required for the hammer, although pins may be added to the sides of one of the wheels for this purpose.

There is usually a lifting cam to move the rack stop clear of the teeth of the rack as it is advanced during striking. A warning stop pin is usually found in the rim of the last wheel in the train.

2. Cam and Lever System:

The operation of the rack and snail strike system is controlled by interacting levers and cams. Their function is to start and stop the striking action and to define the number of strikes at each hour.

a. The Rack:

This is a steel segment of a circle with a number of fine teeth cut in its outer edge. At one end of the curved segment there is an arm mounted on a pivot pin.

A second arm, which is part of the same piece, has a pin at its end, or is bent at a right angle to form a tab. This arm is called the Rack Tail. The pin or tab is located so that, when the rack is dropped, it comes to rest on a surface of the snail.

b. The Cam and Gathering Pin or Gathering Pallet:

We have shown a typical American cam with a gathering pin. Gathering pallets, as found on many clocks, particularly those of European origin, are in fact tiny cams and perform the same function.

The cam and gathering pin or gathering pallet must perform two functions; to raise the rack stop clear of the rack teeth, and to advance the rack, one tooth at a time. Most American clocks and many foreign clocks provided with rack and snail striking employ a cam and gathering pin. Older English and European clocks commonly use a gathering pallet.

(1) Cam and Gathering Pin:

Our illustrations show this device.

On the end of an arbor which projects through the face of the front plate is mounted a cam with a pin. This cam rotates as the arbor turns. The cam raises the rack stop clear of a tooth while the pin engages with a tooth of the rack to move it upward. This pin is called the gathering pin.

Before the pin clears the tip of the

rack tooth, the cam lowers the rack stop between teeth, so that the rack is held until the pin again moves around and gathers in (comes in contact with) the next tooth.

(2) Gathering Pallet:

A gathering pallet is a small wedge shaped steel cam. It is mounted on an arbor projecting through the front plate. It has a point at one end and a wider, rounded face at the other. It performs the same functions as the cam and pin described above.

The point of the gathering pallet moves the rack, one tooth at a time, with each revolution, while the wide end simultaneously lifts and lowers the rack stop lever.

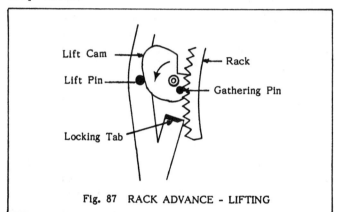

Fig. 87   RACK ADVANCE - LIFTING

c.   Rack Advance Operation:

(1) Lifting:

As the lift cam rotates, as shown by the arrow, the lift pin riding on its outer surface moves the locking tab of the rack stop lever clear of the next tooth in the rack. At the same time, the gathering pin on the cam starts to engage the next tooth.

(2) Moving:

The pin on the rotating cam pushes the tooth with which it is engaged and moves the rack upward. The cam holds the locking tab well clear of the top of the next tooth.

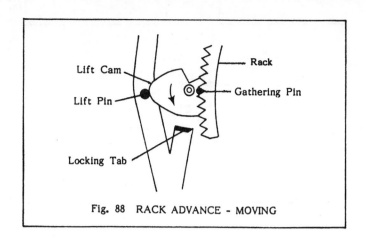

Fig. 88   RACK ADVANCE - MOVING

(3) Locking:

As the gathering pin begins to clear the tooth it has been pushing, the cam has rotated the lift pin rests on a shallow face, lowering the locking tab into position to lock the next tooth.

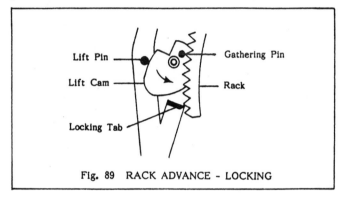

Fig. 89   RACK ADVANCE - LOCKING

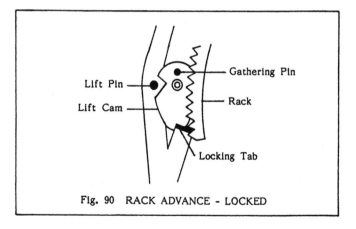

Fig. 90   RACK ADVANCE - LOCKED

(4) Locked:

The gathering pin is now out of contact with a tooth of the rack. While the cam rotates to bring the pin into position to gather the next tooth, the locking tab remains in position against a tooth, to prevent the rack from falling.

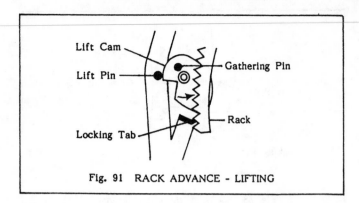

Fig. 91 RACK ADVANCE - LIFTING

(5) Lifting:

As the cam continues to rotate, the cam begins to raise the lift pin, starting to move the locking tab away from the rack tooth. The tab will retain the tooth until the gathering pin comes in contact with the next tooth, as the cycle begins again. (Fig. 87)

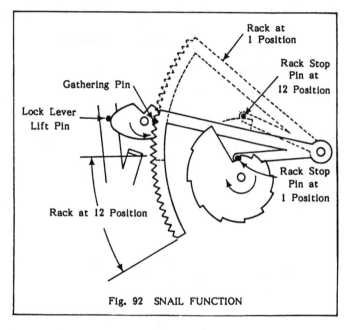

Fig. 92 SNAIL FUNCTION

d. The Snail:

Mounted on the cannon (hour hand) tube, the snail is actually a form of cam with twelve progressively deeper segments. It makes one revolution in twelve hours, so the segments progressively move into the position at which the rack tail falls at each hour.

When the warning cycle starts, the rack locking tab moves away from the end of the rack, allowing it to drop. The rack stop pin on the tail of the rack lever comes into contact with a segment of

the snail, stopping its movement.

The highest segment on the snail stops the dropping of the rack after only one of its teeth has passed the locking tab. The next segment allows the rack to move two teeth, and so on, until the deepest segment is reached, when twelve teeth will pass. The solid lines in Fig. 92 show the rack in this position.

The particular segment that is in position when the rack tail is dropped determines the number of teeth which must be lifted by the gathering pallet and, thus, the number of times the strike lever will be activated.

The dotted lines in Fig. 92 show the rack and part of the snail after the rack has dropped and been stopped on the outermost segment of the snail. The dotted line on the rack near the rack locking tab indicates the end of the rack, with only one tooth to be moved up.

e. The Rack Stop - Warning Lock Lever:

As the gathering pallet moves away from the tooth it has advanced, a pin or hook on the rack stop lever drops between teeth to hold the rack until the gathering pallet returns to pick up the next tooth.

When the last tooth of the rack is passed, the rack stop drops down under the end of the rack and the warning lock tab on another arm of the lever moves into position to stop further movement of the train by interfering with a pin on one of the strike train wheels.

f. The Lift-Lock Lever:

This lever has three arms. One has a warning lock tab at its end. The middle one has a pin for moving the warning-rack stop lever. The third has a tab which rides on the lift cam.

(1) The tab of the first arm rides on the lift cam, which raises and lowers it.

(2) As the cam raises this lever, the lift pin on the end of the second arm, comes in contact with the warning-rack stop lever and moves it to release the lock pin and

allow the rack to drop.

(3) The third arm is moved into position so that the tab at its end is in the way of the pin on the wheel, providing the warning lock, cocking the movement, ready for striking. When the tip of the cam passes, the lever drops, the movement is freed and striking begins.

4. Operation of the Rack and Snail System:

All the levers of this system are moved, directly or indirectly, by the lift cam on the minute arbor, or by the cam and pin mounted on the end of an arbor of the strike train.

The functions performed are the same as those required of the count wheel system, lock, warning and striking.

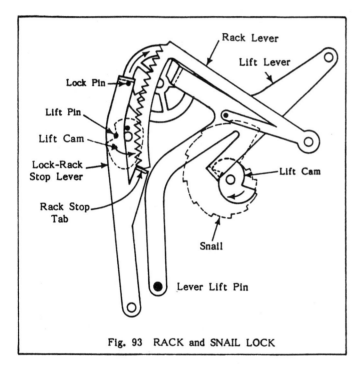

Fig. 93 RACK and SNAIL LOCK

a. Locking:

Except when actually counting the hours, the strike train is locked as shown. The tab at the end of the lower arm of the lift lever has just dropped off the lobe of the cam on the minute arbor. The lever lift pin on the center arm has moved away from the locking-rack stop lever, allowing it to move into position to stop the rack from falling and the tab at its end has moved in to intersect the

lock pin on the wheel, stopping or locking the train.

Note that the pin on the rack tail is shown above the highest segment of the snail.

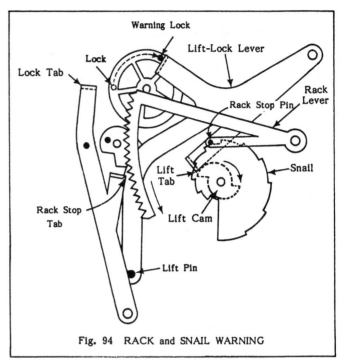

Fig. 94 RACK and SNAIL WARNING

b. Warning:

The cam on the minute arbor has raised the rack tail enough to cause the lift pin on the end of the lock lift lever to push the lock-rack stop lever and its tabs clear of the rack and the lock pin on the wheel. It has also moved the lift lock tab into position to interfere with the pin on the wheel.

The rack has fallen, until stopped by the rack stop pin as it strikes a segment of the snail. The lock tab has released the stop pin, allowing the wheel to turn, until the pin is again stopped by the warning lock tab. The train is now cocked and ready to strike.

c. Running:

The lift cam on the minute arbor has now moved so that the tab on the lift lever has dropped off the point of the cam, simultaneously moving the locking tab away from the pin on the wheel. The strike train has started running. The pin on the

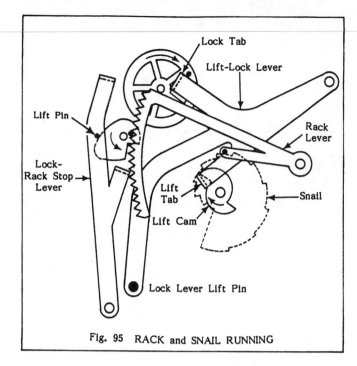

Fig. 95   RACK and SNAIL RUNNING

lift lever has moved away from the rack stop lever, allowing the rack stop tab to come between the teeth of the rack and the lift pin to come into contact with the face of the cam.

Striking proceeds. The cam and the gathering pin will move the rack upward a tooth at a time, until the end of the rack is reached, the rack stop tab falls under the end of the rack and the locking tab moves in front of the pin on the wheel, stopping the train in the locked position.

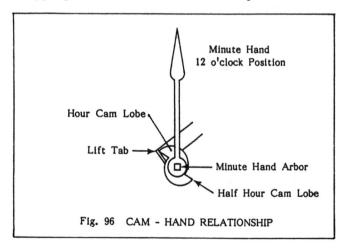

Fig. 96   CAM - HAND RELATIONSHIP

4.   Lifting Cam - Minute Hand Relationship

There is a fixed relationship, in nearly all cases, between the position of the minute hand on its arbor and the lifting cam on the same arbor. The square

or oblong on the end of the arbor fits the mating hole in the hand closely and the cam is firmly attached so that the minute hand points to the figure 12 on the dial at the instant the lift tab falls off the point of the cam lobe.

5.   Variations:

a.   Shape and Location of Parts:

There are many variations in the shapes and locations of the various elements in rack and snail strike trains, but they all function in basically the same manner.   On examination of your movement, by comparing it with the accompanying figures, you will be able to identify and trace the functions of the individual parts described.

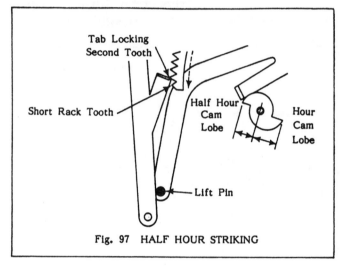

Fig. 97   HALF HOUR STRIKING

b.   Half Hour Striking:

A common means of providing a single strike at the half hour is achieved with only minor changes in the elements of the train.   A second lobe is added to the lifting cam on the minute arbor, diametrically opposite the hour lobe.   The distance from the center to its point is very slightly less than to the tip of the hour lobe.

The first tooth on the rack is very slightly shorter than the other teeth.

When the lift lever is raised by the half hour cam lobe, it moves the lift-lock lever up to release the train.   The rack stop tab is raised only enough to clear the short first tooth on the rack, but not enough to clear the second tooth.

When the lift tab drops off the cam lobe, there is only one tooth to gather before the lock lever again moves under the end of the rack and striking is stopped. Only one strike occurs.

III. STRIKE TRAIN PROBLEMS and SOLUTIONS:

A. FAILURE to STRIKE or to STRIKE PROPERLY:

When a clock fails to strike, or does not strike properly, it is usually concluded that something must be radically wrong with the strike train. This is not always the case.

Many of the problems that develop in strike trains are the same as those in time trains. Those we have discussed in Chapter 8, in which we made no distinction between the two trains.

1. Train Problems:

If the clock fails to strike, it is likely to be due to the same causes that stop time trains. Look for:

a. Dirt on wheels, pinions or in pivots.

b. Broken or seized mainspring.

c. Teeth missing or bent.

d. Pivots or arbors broken or bent.

e. A manual strike lock. Some clocks have a lever, operated from outside the movement, which holds the lock lever, to keep the clock from striking.

2. Synchronization Problems:

Unlike the time train, the strike train performs several independent, but related functions, in a predetermined sequence. To do this, each wheel in the train must be in a certain position with respect to other wheels that perform related functions. For example, in a count wheel system, when the end of the count lever is in a slot of the count wheel, the slot in the locking cam

attached to the third wheel must be in exactly the right position to receive the end of the lock lever.

Since the teeth of one wheel are engaged with those of the pinion of the next, the relationship of wheels to each other is determined at the time of assembly. If that relationship is not right, it is because the movement has been improperly assembled, or there has been such significant wear that wheel and pinion teeth have become disengaged momentarily. In either case disassembly of the movement will be necessary.

Before you decide that you must disassemble the movement, for any reason, you should be thoroughly familiar with the procedure for proper reassembly of the strike train as described in Chapters 10 and 11.

We must assume that all elements were properly synchronized when the clock left the factory. Unless some unknowledgeable workman has disassembled and rearranged the wheels, they should still be in synchronization. A careful check will be necessary to determine if they have been desynchronized.

a. Checking Synchronization-Count Wheel System:

(1) Attach the minute hand to its arbor.

(2) Partially wind the strike train main spring, or, if yours is a weight driven movement, pull on the cord of the strike side winding drum.

(3) Locate the point where the end of the lift lever will be lifted by the cam or pin on the minute arbor and slowly rotate the minute hand in a clockwise direction, until the cam or pin just begins to raise the lift lever.

(4) Now locate the end of the count lever, which should be in a slot of the count wheel, and the end of the lock lever, which should be in the lock position in the slot of the cam.

(5) Gently touch the teeth of the last wheel in the strike train. (The wheel

that engages with the pinion of the fly arbor.) This will act as a brake, allowing you to start and stop the strike train, after it has been released.

(6) Slowly advance the minute hand until you see the count lever begin to rise out of the deep slot in the count wheel in which it has rested and the lock lever rise out of its slot in the cam.

Watch for the warning phase, when the warning lever stops the rotation of the fly.

Carefully continue advancing the minute hand until the warning lever drops clear of the fly. At this point, if it does not point to 12, relocate the minute hand on its arbor so that it does. It will then be synchronized with the strike.

b. Rack and Snail System -Checking Synchronization:

(1) Attach the minute hand to its arbor.

(2) Partly wind the strike train main spring, or if you have a weight driven movement, wind a few turns, then pull on the cord of the strike train drum.

(3) Locate the point where the end of the lift lever will be lifted by the cam, or pin on the minute arbor and slowly rotate the minute hand in a clockwise direction, until the cam or pin just begins to raise the lift lever.

(4) Gently place the tip of a finger on the tips of the teeth of the last wheel in the strike train. (The wheel that engages with the teeth of the pinion of the fly arbor.) This will act as a brake,

allowing you to start and stop the motion of the strike train, after it has been released.

(5) Locate the rack stop lever, which should be below the end of the rack and the pin or tip of the gathering pallet which may or may not be engaged with a tooth of the rack.

(6) Locate the point where the stop lever is engaged with the stop pin.

(7) Lift your finger from the wheel so that it is just barely in contact and ready to slow or stop the motion of the train. Slowly rotate the minute hand and observe these actions which should occur several minutes before the minute hand reaches the hour position:

(a) The rack stop lever will be raised by the lift lever and the rack allowed to drop to a segment of the snail. At the same time, the locking lever will be lifted away from the locking pin, allowing the train to start moving.

(b) The warning lever will quickly move into position to again stop the movement by engaging with the warning pin. The strike train is now locked in the warning position and is ready to strike as soon as the lift lever drops off the cam on the minute arbor.

(8) Continue slowly rotating the minute hand, prepared to stop the instant the warning lever is released and striking begins. Relocate the hour hand on the minute arbor, so that it points to 12. The hand is now synchronized with the strike.

Chapter 10

BASIC MOVEMENT DISASSEMBLY,

CLEANING AND REASSEMBLY

The prospect of disassembling and reassembling a clock movement for the first time can be awesome.

In itself, tearing a movement apart is not at all difficult. However, doing so in such a way as to avoid damage to its parts and to facilitate proper reassembly requires some study, planning, and use of a simple, methodical procedure.

Disassembly, and reassembly, requires only ordinary mechanical dexterity and a reasonably steady hand. No fancy tools or equipment are necessary, but you may want to acquire a few inexpensive tools you may not have.

I. REASONS for DISASSEMBLY and REASSEMBLY:

A. For LEARNING:

If you are interested in learning about clock repair in general, we suggest disassembly, cleaning and reassembly of a working movement, not in need of repair, as an important and most worthwhile exercise.

Using a working movement, you will be able to concentrate on the procedures involved, without concern for correcting defects. When reassembled, the movement should again run as well or better than it did before you disassembled it.

1. Selecting a Movement:

a. Weight Driven, 30 Hour:

As a first project, we recommend that you use a 30 Hour, weight driven movement, as found in American ogee or half-column shelf clocks.

Because the weights are easily removed and there are no main springs, there is no stored power to contend with. These movements have both time and strike trains, with count wheel strike system, in their simplest forms.

It is one of these movements whose disassembly and reassembly we will describe in some detail in this Chapter.

If you do not already have one of these movements, they can be obtained at relatively nominal cost at most NAWCC Regional Meetings and some Chapter meetings, from a clock dealer, or at a flea market or antique show.

b. Spring Driven, 30 Hour:

Second choice for your first exercise in disassembly and reassembly would be an American 30 Hour spring driven movement, as found in many small shelf or "cottage" clocks.

Because they are required to run the movement for only 30 hours, the springs are much smaller and less powerful than those found in 8 Day movements and are easier to deal with in disassembly and reassembly.

Ideally, you would use this type of movement as your second practice project, after first working with a weight driven movement.

Chapter 11 deals with the assembly and disassembly of spring driven movements. If this is your first project, however, you should finish reading this Chapter, which contains basic information applicable to movements of both types. It will be helpful to have the spring driven movement at hand for comparison, as you read.

B. To ACCOMPLISH REPAIR:

If you have a movement that doesn't run, after you have gone through all the steps covered in preceding chapters, and you have determined that you can perform the repair or replacement required, you may want to proceed with disassembly.

See Chapter 12 and the Index for some specific repair procedures you may be able to accomplish.

It is also possible that you may want to perform disassembly in order to remove a part to be worked on by a more skilled, or better equipped, repairman.

CAUTION:

It is strongly recommended that you read through the procedures involved in both disassembly and reassembly, with your movement available for comparison, BEFORE you begin the work. If yours is a spring driven movement, carefully read Chapter 11.

When you feel you understand the simple procedures involved, you can proceed with confidence. We strongly recommend that you follow these procedures in the sequence described, using the book as a guide.

1. Simple Repairs the Beginner can Expect to Make:

a. Cleaning and Lubricating:

Since accumulations of dirt can prevent a movement from operating, cleaning is a form of repair. As indicated earlier, thorough cleaning can

only be properly accomplished after disassembly of the movement. This is especially true with respect to springs.

b. Spring Reattachment or Replacement:

In nearly all cases, the movement must be disassembled in order to replace broken springs, or to reattach those which may have come unhooked from the pin on the winding arbor, or on the inside of a barrel.

Some clocks of comparatively recent manufacture, however, have separate removable plates attached to the movement so that mainsprings can be removed without complete disassembly of the movement.

CAUTION:

Before proceeding with disassembly in order to replace a spring, it is strongly suggested that you carefully read the section in Chapter 11 dealing with springs.

c. Parts Replacement or Repair by Others:

Before disassembling the movement, to replace a defective part, you would be well advised to make sure that the part or parts you need are available, or that you have located someone able to do the necessary repair work. If a long time goes by between disassembly and reassembly, particularly with your first experience, you may find it more difficult to remember details that make reassembly easier.

II. DISASSEMBLY of a 30 HOUR WEIGHT MOVEMENT:

The procedure we will describe and illustrate here will be based on a brass plate 30 Hour weight driven movement made by Ansonia. The steps covered, however, will be found generally applicable to most other 30 Hour weight driven brass and wood movements, as well.

It is assumed that the hands, pendulum and weights have been removed, as discussed in earlier Chapters and that the movement has been taken from the case.

A. GETTING READY:

1. The Work Place:

A good, comfortable place to work is important. It should have a clean, flat work surface, such as a table, desk, or bench, that is solid and not subject to movement if someone happens to bump it.

Good lighting is essential, since you will be working with some very small parts. Some form of concentrated lighting such as a flexible desk lamp, or high-intensity lamp will be most helpful.

2. Tools and Accessory Items:

The only tools you will need are among those listed in Chapter 6.

a. A Small Container for Loose Parts:

It is important to keep loose parts, such as bobs, nuts and screws, in some sort of container, so they won't be misplaced or lost.

We have found one of the best such containers is a plastic sandwich box, with a snap-on lid. It is large enough to hold most hands, bobs and levers, as well as small parts like nuts, washers and pins.

Another good container is a plastic one pound margarine or similar food container, with a lid.

b. A Plastic Foam Block:

This is a most useful means of storing arbors and wheels in proper sequence, as they are removed from the movement.

You may find suitable material used as packing for many household items, such as small appliances. It is also available for purchase in blocks at many discount or variety stores. You will need a piece at least 4" wide and 8" long. It should be at least 3/4" in thickness.

B. STUDYING the MOVEMENT:

Take a few minutes to study the arrangement of all the parts in the movement. We will specifically describe and illustrate an Ansonia weight driven movement. Your movement may vary from this one in some details, but will be similar.

1. Front Plate Attachment:

Study the front plate and note the location of the pillar ends that extend through the plate and are, or should be, held in place by taper pins, or nuts. There are usually four pillars, the lower two usually near the corners of the plate, while the other two may be randomly located near the top.

2. The Time Train: Fig. 98

As shown in Fig. 98, the time train is the series of wheels and pinions, usually located at the right side as you face the front of the movement. As in this case, a 30 Hour movement usually has only three arbors in the time train.

Look carefully at the wheels that constitute the time train and the relationship between the wheel on one arbor and the pinion of the next, with which it engages.

Note the relative position of wheel, pinion and other elements on each arbor. Some pinions will be positioned on the arbor between the wheel and the front plate, others may be between the wheel and the back plate.

a. Motion Work: Fig. 99

The time train is a continuous series of large wheels mating with small pinions that turn successive arbors at progressively increasing rates of speed. At the end of this train is the escape wheel, whose speed is controlled by the pendulum.

Another set of wheels and pinions is necessary to provide a visual indication of the passage of time, as measured by the time train. This is called the motion work. Its sole function is to turn the minute and hour hands.

Near the center of the front plate you will see the projecting minute arbor with its rectangular end on which the minute

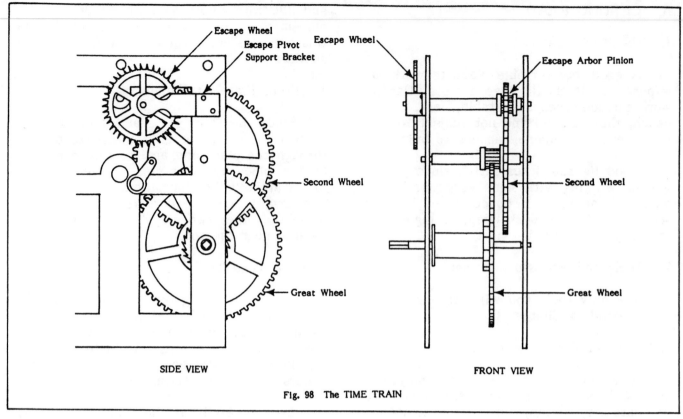

Fig. 98 The TIME TRAIN

SIDE VIEW   FRONT VIEW

hand was mounted. Surrounding the minute arbor is the slightly shorter cannon tube which carried the hour hand.

Looking between the plates, you will see an arbor just below the minute arbor, on which is mounted a wheel and pinion.

This is the motion work arbor.

The cannon tube fits over the minute arbor and turns independently of it. It has a wheel attached to its inner end which is driven by a pinion on the motion work arbor.

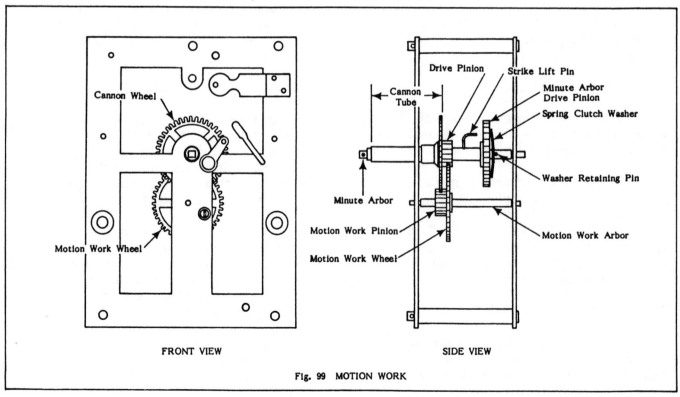

FRONT VIEW   SIDE VIEW

Fig. 99 MOTION WORK

This arrangement results in a speed reduction causing the cannon and hour hand to rotate once in 12 hours, while the minute arbor and hand makes 12 revolutions.

A large pinion on the minute arbor transmits power from the second wheel of the time train to that arbor. If it were firmly secured, it would be impossible to turn the minute hand, which is attached to the minute arbor, to reset the time. To make it possible to turn the arbor, its drive pinion is secured by friction. A spring clutch washer, secured by a pin through the arbor, exerts pressure on the face of the pinion, pushing it against a flange on the arbor. This creates sufficient friction for the train to turn the arbor during its normal operation, while allowing the hand to be turned manually when resetting is necessary. The washer is similar to that shown in Fig. 101.

You will find a bent wire, the strike lift pin, inserted into the minute arbor with its bent end parallel to the arbor. this pin acts to raise the lift lever of the strike train, as it rotates. In some cases, there may be a cam to perform this function, rather than a bent wire.

3. The Strike Train:

At this point, it may be helpful to refer to the section in Chapter 9 which discusses the count wheel strike system to refresh your memory on how it operates.

Usually located on the left side of the movement, as you face the front plate, a 30 Hour strike train commonly has three arbors, two with wheels and the third with a fly governor. The second and third arbors have pinions.

a. The First Arbor:

The first arbor carries the winding drum, the ratchet wheel, the first wheel and the count wheel. The front end extends through the front plate and is squared to fit the winding key.

The first wheel, in addition to transmitting power to the second wheel, also lifts and drops the strike hammer.

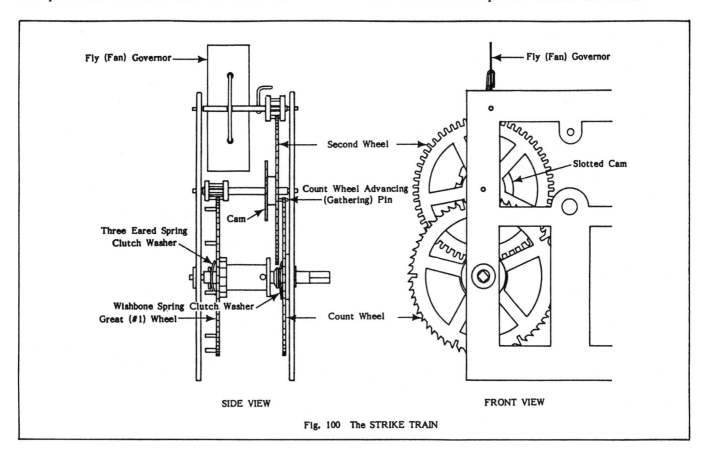

SIDE VIEW

FRONT VIEW

Fig. 100 The STRIKE TRAIN

To accomplish this, a series of equally spaced pins have been inserted in holes and riveted to the rim of the wheel, with the ends projecting toward the back.

In order to be able to wind the cord on the drum, the drum and ratchet wheel must be capable of turning, while the first wheel remains stationary. This is achieved by a three eared friction washer which exerts pressure on the back of the wheel to force it against the face of the winding drum. See Fig. 101.

Near the front end of the first arbor is the count wheel. It also must remain stationary when the arbor is turned during winding. Friction is imposed by the pressure of a wishbone washer as shown in Fig. 102.

a. The Second Arbor:

A pinion is mounted at the back end and the second wheel at the front of this arbor. The pinion is driven by the first wheel and the wheel drives the third, or fly, arbor.

Just in back of the second wheel, it also carries the cam which lifts the count lever and has a slot into which the lock lever drops at the end of a cycle to lock the train.

On the front face of the second wheel is mounted a gathering pin, located so that it engages with the teeth of the count wheel on the first arbor. This pin advances the count wheel one tooth each time the wheel on which it is mounted makes a revolution.

c. The Third Arbor:

This arbor has a pinion at the front end, driven by the second wheel. It also has an L shaped wire which fits through a hole near the pinion and is firmly riveted. This is the warning lock pin. There is a small groove in the arbor, near the back end, into which the center of the tension spring of the fly fits. This keeps the fly from moving out of position laterally on the arbor.

4. Friction Washers:

Friction washers are commonly used in clocks, but are an unfamiliar device to many. A fuller explanation of their function and design may be of value in helping you understand an important detail that makes it possible for clocks to function as they do.

As it happens, the two most commonly used types of friction washers are found in the movement we have described here. Both are made of thin sheet spring brass.

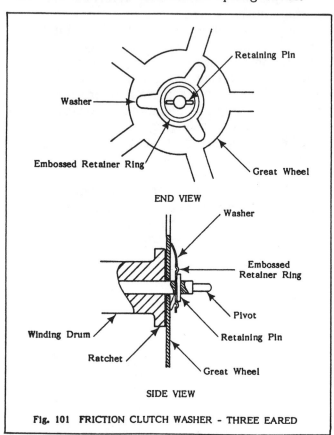

Fig. 101 FRICTION CLUTCH WASHER - THREE EARED

a. Three Eared Friction Washer:

Used in the movement described to provide a clutch or slipping action during winding of both the time and strike train, this type has three tabs, or ears, bent downward to provide a spring action.

The top of the washer, with a hole in its center, is embossed to form a raised ring. The hole in the washer fits over the winding arbor, in which there is a small hole for a retaining pin.

After the first, or great wheel has been placed on the arbor, the washer is put on. The tabs are designed so that the top of the washer is outside the hole in the arbor at this point. Pressure must be applied to bend the ears sufficiently so that the retaining pin can be inserted.

When the pin is in place, the ears of the washer exert a constant pressure on the face of the wheel.

The raised ring in the top of the washer prevents the retaining pin from coming out of its hole.

When the arbor is turned, during winding, the wheel and washer remain stationary, while the winding drum and the retaining pin turn. During normal operation, the pressure exerted by the washer is sufficient to transmit necessary power to the train.

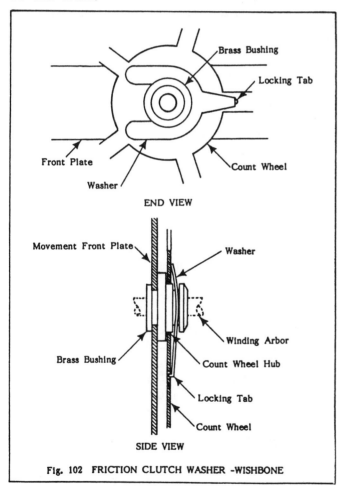

END VIEW

SIDE VIEW

Fig. 102 FRICTION CLUTCH WASHER -WISHBONE

b. Wishbone Friction Washer:

This washer derives its name from its shape, resembling a wishbone. It has two wide arms at one end and a single narrower tapered arm, with a small tab at its end, at the other. It is curved from one end to the other. In the movement described it is used to provide a clutch action for the count wheel.

A brass bushing is attached to the front plate of the movement. Inside the plate, it has a short section on which the hole in the count wheel fits and a shoulder against which the wheel bears.

There is groove in the bushing near its inside end. The two arms of the washer slide into this groove.

The tab at the end of the third arm fits into a hole in a spoke of the count wheel, to lock it in position.

The curvature of the washer and the location of the groove in the bushing are such that pressure must be exerted to place it in the groove. When in place, the three arms of the washer exert force against the face of the count wheel to keep it in position.

This arrangement makes it possible for the winding arbor to be turned, while the count wheel remains stationary.

5. The Strike Lever System:

Locate the two lever arbors and the wheels with which they interact. By pulling on the cord and lifting the count lever from the slot in which it rests in the count wheel, let the train operate until it stops when the count lever drops into a slot. Notice the relationship of the levers mounted on the two arbors.

Also note the way the bent end of the lock lever fits into the slot on the cam on the second wheel arbor, when the count lever is in a deep slot on the count wheel.

It is a good idea to make a sketch of the position of these elements in the plate and their relationship to each other, for your movement, if it differs from ours.

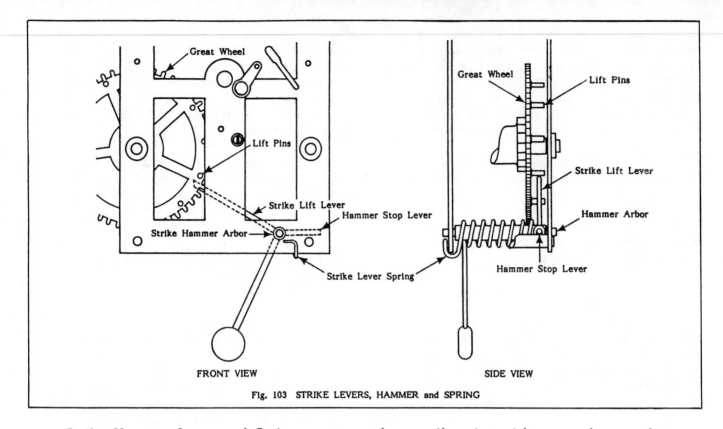

Fig. 103 STRIKE LEVERS, HAMMER and SPRING

FRONT VIEW   SIDE VIEW

a. Strike Hammer Lever and Spring:

Near the bottom of the plates, you will find the strike hammer lever arbor. Three arms, or levers, project from it. One, the hammer lever, has a heavy brass disc, the hammer, at its end.

The strike lift lever extends upward and to the left with its end just past the rim of the first wheel of the time train. This lever is raised by the pins on the wheel and dropped as the pins pass its tip.

Notice that the end of the lift lever lies between and is not in contact with any of the lift pins on the first wheel, when the movement is locked.

A third, shorter lever extends to the right above and just past the movement pillar. This is the hammer stop lever, or hammer tail. When the hammer is dropped, this tail strikes the pillar, stopping its motion. When this happens, the weight of the hammer causes its arm to flex, very slightly, so that it strikes the gong, but immediately rises clear, leaving the gong free to vibrate.

Wrapped around the hammer arbor will

be a coil spring with one end secured to the strike lifting lever and the other hooked over the edge of the movement. Its purpose is to accelerate the movement of the hammer arm to give a sharp blow to the gong. This spring must be unhooked from the front plate before disassembly of the movement is started.

b. Activating Levers: Fig. 104

In Chapter 9, we explained the operation of the levers in count wheel systems. Fig. 104 shows their relative position in the movement.

All levers are shown in their at rest, or locked, position. Note particularly the side view for the location of the levers along the length of their arbors.

Reference to this figure when you start reassembling a movement will make clear where each lever should be.

c. Count Lever Lift Wire:

Near the outer end of the count lever, you will probably find a small loop into which a loop on the end of a straight piece of fine wire is inserted. This wire hangs down, along the outer edge of the

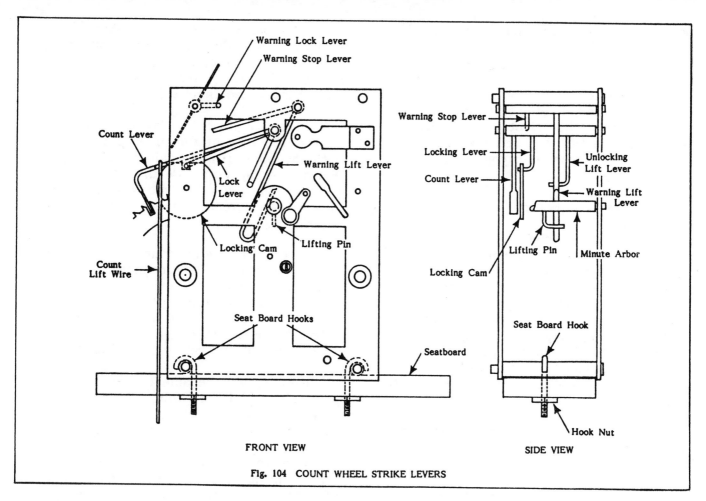

FRONT VIEW   SIDE VIEW

Fig. 104 COUNT WHEEL STRIKE LEVERS

front plate and then usually passes through a small staple in the face of the seat board.

The purpose of the lift wire is to permit manual lifting of the strike lever, which causes the clock to strike. By lifting the wire and allowing the clock to strike, and repeating until the number of strikes matches the time shown by the hour hand, striking can be synchronized without moving the hour hand.

Often this wire is missing. Since the hour hand of nearly all count wheel strike clocks is a friction fit on the cannon tube and can be safely moved to match the hour struck, a lift wire is not essential to satisfactory operation of the clock.

6. Seat Board and Seat Board Hooks:

Notice that the seat board fits between the front and back plates of the movement and is in contact with the bottom of the two bottom movement pillars. It will have a section cut out where the

hammer lever passes through it.

The seat board is attached to the movement and held securely by the seat board hooks that hook over the bottom pillars at their top ends and have a round or square thin brass nut under the seat board at the opposite ends.

Remove the seat board before starting disassembly.

C. BEGINNING DISASSEMBLY:

Having carefully studied the movement, knowing that you are about to disassemble it, you should feel more confident. You will probably want to refer to the illustrations from time to time as you proceed.

Before starting, notice that the pivots of each arbor project beyond the faces of both back and front plates by a noticeable amount. Each of them can be moved back and forth, a little bit, between the plates. This clearance is

103

known as "end shake" and is normal.

1. Remove Anchor (Verge) Assembly:

Slightly lift and rotate the flat end of the "L" shaped retainer spring that rests on the anchor pivot pin, so that it is clear, then lift the anchor assembly off the pin. Place it in your parts container.

2. Remove Seat Board:

If the nuts holding the seat board hooks are round, use special care in removing them. Holding the sides of your flat nose pliers flat against the bottom face of the seat board, grip the nut securely and rotate counter-clockwise. Release and repeat, until the nut can be removed with the finger tips. When the hooks are loosened enough to clear the bottom pillars, rotate them and remove the board.

As you take the seat board away, be careful to clear the strike wire, if there is one, from its staple.

3. Unhook the Strike Lever Spring:

Place the point of a small screwdriver in the hook of this spring, where it goes over the edge of the front plate and gently move it down and in, until it is clear of the plate, then release it.

4. Remove Front Plate:

You are now ready to lay the movement on its back to remove the front plate.

If you place it on a hard, flat surface, pressure will be exerted on the projecting ends of the arbor pivots. At this point, if you have them, attach assembly clamps (Fig. 42) to the back plate, one near each bottom pillar and the third near the center at the top.

If you don't have clamps, simply fold a dish towel several times to provide a pad, which will serve to absorb pressure from the projecting pivots and cushion the movement during disassembly.

If you have modified your flat nose

pliers for use with taper pins, (See Chapter 1) use them now to remove the pins from the pillar projections. If you have not modified your pliers, it will be necessary to use their sides.

Carefully place one jaw of the pliers against the face of the pillar projection, with the other firmly against the small end of the taper pin. With just a little pressure, the pin will come loose and can be easily removed. Remove the other three taper pins.

CAUTION: Before removing the front plate, notice that the escape wheel is in front of it, while the lantern pinion on the opposite end of its arbor is engaged with the third wheel. If the plate is lifted carelessly, this can cause disturbance of several wheels. (Fig. 98)

Now we will be able to lift the front plate clear, while allowing all pivots, except that for the escape wheel, to remain in position in the back plate.

With the movement on its back and the top of the movement nearest you, gently lift the top end of the top plate. As you do so, notice that the front escape arbor pivot comes clear of its hole, before the plate touches the back of the escape wheel.

A large opening in the plate will allow you to move the escape wheel arbor top pivot clear of its hole, then lift the back pivot and free the pinion so the plate can be lifted clear without serious disruption of the wheels.

Be prepared for the count wheel to come away with the front plate, to which it is attached by a brass bushing and brass split spring washer. (Fig. 100 and Fig. 102)

5. Remove Wheel and Lever Arbor Assemblies:

HAVE YOUR PLASTIC FOAM BLOCK HANDY.

We will use this block to insert arbors, as they are removed, in the same order in which they were installed in the movement. Time train elements should be placed on the right side and those for the strike train on the left. Minute arbor and lever arbors will be at or near the center.

a. Lift Arbors from the Back Plate:

Begin with the wheel which is on top, in this case, the escape wheel arbor. Lift it out, being careful to clear the pinion from the wheel with which it is engaged, as you lift. Then insert the end that was in the back plate firmly into the foam block, in about the same position in which it was located in the movement, near the upper left hand corner.

Since the pivot is very short, you may have to exert enough pressure to recess the pinion in the foam, in order for it to stay in position.

Working down the strike train, next remove the fly arbor, followed by the arbor of the last large wheel in the strike train. Place it in the foam block.

Now, carefully take out the two lever arbors, then the minute arbor and cannon assembly. Locate them in the foam block, in the same relative positions they occupied in the movement.

Continue, until all arbors are removed from the plate and inserted into the foam block.

If your movement does not need thorough cleaning, you may want to skip the next section and go to reassembly. We think, however, you will find it very much worthwhile to read about cleaning anyway.

III. CLEANING a 30 HOUR BRASS WEIGHT DRIVEN MOVEMENT:

A. IDENTIFYING PARTS:

For cleaning we must put all the parts we have so carefully positioned in the foam block separately, but as a group, into a cleaning solution. Look at each of them carefully, as they stand in the foam block.

There are some simple rules for again sorting them into proper order. Usually,

each succeeding wheel is somewhat smaller in diameter than the one preceding it.

Starting with the first wheel, the proper end-for-end position can be determined by matching the wheel of the first arbor to the pinion of the second, etc.

Arbors of the strike train can be distinguished from those of the time train by the special parts needed for the strike function. Check Fig. 98, Fig. 99 and Fig. 100 to verify the sequence.

Levers can be properly positioned by referring to Fig. 103 and Fig. 104.

We suggest that you arrange the arbors in proper sequence in the foam block, before starting reassembly.

A SUGGESTION: In the future, when you are preparing to clean an unfamiliar movement, go through its elements and make notes, as we have done here, to facilitate identification and reassembly after cleaning.

B. CLEANING SOLUTION:

Use this solution for Brass and Steel Movements ONLY.

DO NOT USE for WOOD MOVEMENTS!

If you go very far with clock repair, you will probably hear a great deal about ultra-sonic cleaning, with its expensive equipment and solutions. They have their place in professional shops, but for the occasional repairman, we recommend simple water base ammoniated solutions, which are available in concentrated form from most clock parts supply houses.

Commonly, one quart of concentrate will make a gallon of cleaning solution, which is what you should have.

CAUTION: Carefully read and follow the instructions of the manufacturer for use of their product. In general, such solutions are not highly dangerous, but they usually contain ammonia and should be used with good ventilation.

While these solutions discolor rather startlingly with use, this does not affect their usefulness, until a number of very dirty movements have been cleaned. They can sometimes be re-energized even then, by adding a cup or so of household ammonia.

Cleaning will be accomplished primarily by the chemical action of the solution, with help in brushing away heavier deposits. Like any washing process, this takes time, which may be more or less, depending on the condition of the parts.

The cleaning solution is designed not only to remove oily dirt, but to have a brightening effect on brass parts. Don't expect miracles, though. Most production brass clock parts were not polished when they were produced, so they will not be shiny after cleaning.

1. Mixing and Storing Cleaning Solution:

Mix the concentrate with water, according to the manufacturer's directions, in a gallon glass jug. For safety, be sure to mark it boldly. Store where it is not accessible to children.

C. CLEANING CONTAINER:

The ideal container for this purpose, is a Tupperware or similar container, large enough to accept the plates of the movement you are cleaning. Other good ones are a small 6-quart plastic pail, or gallon plastic ice cream container, with tight lids.

Sealing the container, during cleaning, reduces evaporation of the ammonia and prolongs the life of the solution.

D. PARTS SOAKING and BRUSHING:

The purpose of cleaning is to get rid of dirt and residue that would interfere with proper operation of the clock. Any brightening of the brass is incidental.

1. Place all the movement parts, including front and back plates with assembly clamps removed, in the bottom of

your container. Position them so that flat surfaces are not in contact with each other.

2. Pour the solution over the parts until they are completely covered. Place the lid on the container.

3. Allow to soak for about 20 minutes.

4. During soaking, occasionally lift the container and gently slosh the solution around, to provide a washing action.

5. Open the container and, using rubber or plastic gloves, select a part and, with a small (1/2" to 1") inexpensive nylon paint brush dipped in the solution, lightly scrub it.

Be particularly careful to clean between teeth of both wheels and pinions.

Rinse in clear water to see how cleaning is progressing. By this time, much of the oily dirt should have been removed.

6. Repeat the process again, if the parts are not clean.

Some parts, particularly the plates, may have some areas quite bright, while others are dark and unsightly. After brushing and further soaking, most of the discoloration should disappear.

Seriously discolored parts may be left in most such solutions for several hours. If left too long, however, brass parts begin to take on a reddish cast.

7. After cleaning is complete, return the solution to its storage bottle. Using a plastic funnel, carefully pour the solution back into the bottle. Be careful that no small parts pass into the bottle.

CAUTION: The solution will foam as it is poured and suds will overflow the bottle. Be sure you have it in a sink, where no damage will result.

8. Place your container, with the parts in it, under a faucet and thoroughly flush with HOT water, then shake each part to remove as much water as possible.

9. Carefully dry each part, using a soft lintless cotton cloth. A hair drier, if you have one, will be most helpful for final drying. Use maximum heat and air flow.

Identify and return parts to their proper places in the foam block.

E. FINISH CLEANING:

At this point, carefully inspect the pivots and pivot holes to make sure they are clean and smooth.

1. Pivot Inspection and Cleaning:

Examine each pivot for rust, especially in the area right next to the shoulder that acts as a bearing. If rust is found, grasp a clear portion of the arbor between the thumb and forefinger and spin it back and forth while holding the pivot against very fine steel wool or, as we prefer, between the faces of a small folded piece of crocus cloth. The latter material is coated with fine rouge abrasive and is usually available at auto parts supply stores.

In Chapter 12, we will discuss smoothing and polishing damaged pivots.

2. Pivot Hole Inspection and Cleaning:

Here you will need several round tooth picks. Insert the point of a clean toothpick into each pivot hole, first from the front of the plate, then from the back, and rotate several times, with the fingers. Repeat with a clean tooth pick until all dirt is removed and the pick comes out of the hole clean.

Be sure to clean pivot holes in both front and back plates.

IV. REASSEMBLING a 30 HOUR WEIGHT DRIVEN MOVEMENT:

A. POSITION BACK PLATE:

Reassembly will be much easier, if the movement is positioned at, or slightly above eye level on the table or bench, when you are seated. Any means of doing so that is stable and provides adequate support will do. A pile of books, a corrugated box, or a coffee can of suitable size may be used.

Attach assembly clamps to the back plate, or place it on a folded dish towel, so that pivot ends can go beyond the back face of the plate. This will not be necessary if you use a hollow support, like a coffee can.

B. LOCATE ARBORS in BACK PLATE:

With the back plate in front of you and the foam block with your arbors arranged in sequence handy, you are ready to begin.

The arbors, when placed in the proper pivot holes, should fit well enough to stand reasonably upright, when unsupported. If the holes are worn so badly that some of them won't stand up, they should be rebushed. A simple rebushing procedure is described in Chapter 12.

If you are not prepared to perform rebushing, try leaning the parts toward each other for support, until the front pivots can be inserted into the front plate.

Some unsteadiness is to be expected and you will have to be patient. Even fairly experienced repairmen have been known to utter an oath during reassembly, so be patient and keep trying.

Insert pivots of arbors in pivot holes in the back plate, in the following sequence. MAKE SURE YOU HAVE THE RIGHT ARBOR, in each case and CHECK for PROPER ENGAGEMENT of WHEELS and PINIONS, as you proceed.

1. Second Wheel Arbor, Time (Right) Side.

The wheel will be just above the back plate.

2. First (Winding) Wheel Arbor, Time Side:

This wheel lies just on top of the second wheel. It is the one that winds clockwise.

NOTE: Most winding drums have a hole for the cord which runs diagonally from the face of the drum to the outer face of the flange. If this is the case, the cord should be inserted before the arbor is placed in the back plate.

a. Insert the cord through the hole in the face of the winding drum. Tie a knot near the end and trim close to the knot, then pull the cord so the knot is inside the enlarged hole in the end of the drum.

b. If you want to use new cord, we recommend braided nylon fishing line, 90 pound test. Measure against the old cord and cut about 6" longer. Cut to final length after the movement is installed in case and the cord has been directed over pulleys in the case top.

Make a simple knot at one end of the cord, then cut the cord close to the knot. The short end of the cord should now be fused to keep it from unraveling and to secure the knot.

The slow application of heat, from a match or cigarette lighter, with the flame at right angles to the cord will melt the end of the cord. A small ball of nylon will be formed.

Too much heat will start the cord burning. Use a knife blade or similar instrument, pressing the flat side to the melted ball, to snuff out the fire and flatten the ball. DO NOT USE FINGERS, a small but severe burn can result.

3. Minute Wheel Arbor:

This arbor has the minute wheel and a solid pinion. The cannon tube and pinion should not be on it at this point. It fits into the top one of the two holes in the center of the back plate.

4. First Wheel Arbor, Strike (left) Side:

Attach the cord to the drum, as we described in 2 above.

This arbor is the one that winds counter-clockwise and has pins on the back

of the rim of the wheel.

5. Lift Lever Arbor:

The lift lever has a large loop at its end, the open side of which will rest on the minute arbor. The pivot of this arbor fits into the top hole near the center of the back plate.

6. Count Lever Arbor:

This arbor fits into the hole slightly below and to the left of the lift lever arbor. Its three arms lie between the two of the lift lever arbor.

7. Hour Wheel (Motion Work) Arbor:

This arbor, with its wheel and pinion up, now goes into the hole just below the minute arbor, in the center of the back plate.

8. Second Wheel Arbor, Strike Side:

This last, large wheel arbor fits, with its pinion near the back plate, into a hole just above the rim of the first wheel, with which the pinion engages.

9. Count Lever Arbor:

IMPORTANT: The count lever must now be placed so that the flat end rests in one of the deep slots of the count wheel and the bent end of the lock lever is in the bottom of a slot in the cam on the second wheel arbor.

It may be necessary to disengage wheels and pinions in order to achieve this vital relationship.

9. Fly Arbor:

This arbor fits in the topmost hole on the left side of the plate, with the fly near the back plate and its lantern pinion engaging with the teeth of the second wheel.

10. Strike Hammer Arbor:

Locate this arbor in its hole at the bottom of the plate, making sure that the tail is between the winding wheel and the back plate and between, but well clear of, two pins on that wheel.

B.   POSITION FRONT PLATE, COUNT WHEEL and ESCAPE WHEEL:

Notice that the minute arbor is the longest one and that the winding arbors are next longest. Also note that the ends of the ends of the pillars are slightly longer than the rest of the arbors.

Take the front plate, with the count wheel attached to it by its spring washer and position it carefully just above the partially assembled movement.

1.   Locate the large hole in the center of the plate just over the minute arbor, and gently lower it over the end of that arbor.

2.   Locate the two winding arbors and lower the plate until it is in contact with them, then, with the tip of a finger position them one at a time so that they move into their holes in the plate. Lower the plate carefully.

3.   Position the two holes in the lower corners of the plate over the ends of the bottom pillars and lower the plate over them.

4.   Guide the pivot of the strike lever arbor into its hole.

The plate will now rest on the tips of some of the pivots which will not have moved into their pivot holes.

5.   LIGHTLY, with just finger tip pressure, insert a taper pin in the hole in each of the bottom pillars. This will hold the front plate in alignment, yet allow you to lift it slightly, while you position the rest if the pivots in their holes in the front plate.

6.   Insert the pinion end of the escape wheel arbor through the opening in the front plate and position the pivot in the hole in the back plate, working the wheel under the support bracket on the front plate.

TO FACILITATE LOCATING PIVOTS in holes in the top plate, it is a good idea to locate your light source so that it reflects brightly off the under surface of the top plate, making the pivot holes easier to see. Use tweezers, or the pivot locating tool described in Chapter 12, (Fig. 123) to move the pivots into their holes.

7.   While applying slight pressure at the top edge of the front plate, work the hour wheel pinion into its hole in the front plate. Continue this with each of the remaining pivots working upward toward the top of the plate, until all are in position.

8.   Gently move the top plate down so that it bottoms on the shoulder of the tops of the pillars. It is sometimes necessary to move the tops of the pillars slightly to accomplish this.

9.   With the tip of a finger, move the second wheel in each train, to be sure all wheels and pinions are properly engaged and power is transmitted from one to the other, causing them to turn smoothly.

10. Insert taper pins in the top pillar holes and tighten all four pins with pliers. If your movement plate is secured with nuts, finger tighten them, then secure with a wrench. Remember that these small nuts, made of brass, are easily stripped, so use only a little force.

11. Rehook the strike lever spring over the edge of the front plate.

12. Check the action of both trains by applying a little force, with a finger tip, to each main wheel.

Especially check the function of the time train. While applying pressure to the first or second wheel, making sure you are turning it in the right direction, lift the count lever. Observe the function of all elements during striking and locking. If the entire train does not perform properly, recheck the position of each lever, the cam and the count wheel, comparing it with Fig. 104. Make any necessary correction.

13. Lightly lubricate all pivots at the front and back plates using CLOCK OIL ONLY. DO NOT USE household oil, motor oil, or spray lubricants. Dip the tip of a tooth pick into the oil and carefully touch each pivot. Refer to Fig. 80.

14. Reattach the seat board.

C.   INSTALL the MOVEMENT in its CASE:

1.   Reverse the procedure used to remove the movement.

2.   Run the weight cords over the pulleys in the top of the case.

3.   Place the weights in position on the bottom of the case and tie S hooks on the cord at a point about 1" below the top of the weight. Cut off excess cord and fuse the end as described earlier.

4.   Using the oiling technique described earlier, oil the anchor pivot pin and the anchor pallet contact faces. Put the anchor and crutch assembly in place on its pivot pin and move the retainer over it, so that it is in contact with the pin.

5.   Run the pendulum rod through the loop in the crutch and insert the suspension spring in the slot of its support. Pull down slightly, to make sure it is firmly in position.

6.   Hook the pendulum bob to the bottom of the pendulum rod.

7.   Check to make sure that, with the case level, from front to back and from side to side, the pendulum rod is near the center of the crutch loop. If it is not,

carefully bend the crutch to bring it into proper position.

8.   Check to see that the crutch loop is perpendicular to the face of the movement. If it is not, use pliers to adjust it.

D.   TEST RUN the MOVEMENT:

1.   Wind the weights up so they are at least an inch above the bottom of the case.

2.   Set the pendulum in motion and check to see that the escape wheel is being released by the pallets. If it is not, adjust the crutch as described in Chapter 5. (See Fig. 30)

3.   Lift the count lever and observe the striking function. It is possible that the hammer lever has been bent during handling and does not properly strike the bell or gong. Carefully bend the lever so that the hammer is centered on the face of the gong and comes to rest about 1/16" above it.

SUMMARY

After you have completed reassembly of your movement, we suggest that you take a few moments to thoughtfully review this experience. Do you now better understand how a clock movement works? Do you understand why over-oiling is bad for clock movements? Do you now know the proper way to lubricate movements? Were you able to make some simple repairs, without the need for sophisticated tools?

You may want to review the text, with your movement in hand, to help fix procedures in your memory.

Chapter 11

DISASSEMBLING, CLEANING and REASSEMBLING

An 8 DAY SPRING DRIVEN MOVEMENT

With Count Wheel Strike System

With the exception of the main springs, which replace the weights, the procedure for disassembling and reassembling spring driven movements is essentially the same as for weight driven ones. It is suggested that you review Chapter 10 before proceeding.

30 HOUR MOVEMENTS

As we indicated in Chapter 10, if this is your first experience with the disassembling and reassembling of a clock movement, a 30 Hour spring driven movement from a small American shelf (Cottage) clock is recommended. Since these movements are similar to the weight driven movement discussed in Chapter 10, except for the power source, study Section II of this chapter to learn how to immobilize the springs, then follow directions given in Chapter 10.

8 DAY MOVEMENTS

Movements designed to run and strike for 8 days, seven times longer than the 30 hour type, must store more power, yet release it at the same rate as those of shorter duration. To accomplish this, heavier springs are used and additional arbors, wheels and pinions are introduced.

In this Chapter, we have chosen to describe and illustrate a New Haven Clock Co. movement from their No. 311 shelf clock. While similar, in most respects,

to movements made by many manufacturers during the late 1800's and early 1900's, this one has some comparatively unusual features. Notably, the strike train main spring is much smaller than that employed for the time train. Also, there are five pillars between the plates, rather than four. The fifth one is at the center of the movement, just below the minute arbor.

The basic principles covered here will apply to nearly all spring driven movements with count wheel strike systems.

A step by step procedure check list for reassembly of this movement will be found at the end of this Chapter.

I. PROCEDURES for DISASSEMBLY:

A. EXAMINING the MOVEMENT:

Before starting disassembly, it is essential that you carefully study your movement, comparing its details with those of the movement we will illustrate here. Making careful notes, particularly of the location of wheels, cams, pinions, pins and levers, and any unusual features,will avoid confusion when you start reassembly of the movement.

1. The Strike Train:

Note that there are 5 arbors with their wheels and pinions in this train, rather than 3, as in the 30 hour movement

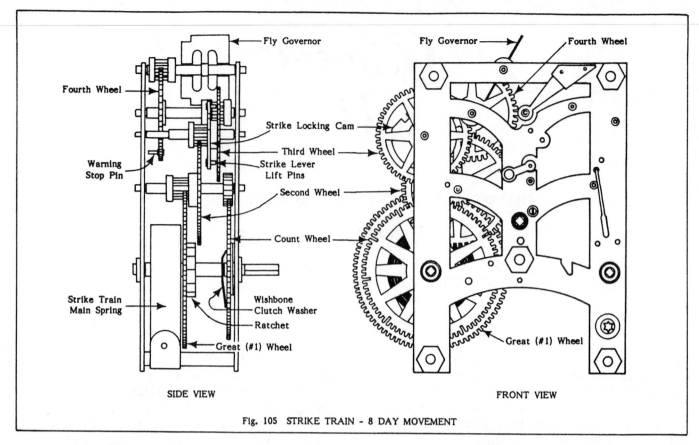

Fly Governor

Fourth Wheel

Strike Locking Cam

Third Wheel

Strike Lever
Lift Pins

Warning
Stop Pin

Second Wheel

Count Wheel

Strike Train
Main Spring

Wishbone
Clutch Washer

Ratchet

Great (#1) Wheel

SIDE VIEW

Fly Governor

Fourth Wheel

Great (#1) Wheel

FRONT VIEW

Fig. 105 STRIKE TRAIN - 8 DAY MOVEMENT

we used in Chapter 10.

a. The strike train main spring is much narrower than the the time train main spring. Compare with Fig. 106. The difference in the springs of this movement is obvious. Less obvious but equally important differences in thickness and length are not uncommon in other movements.

b. The first (great) wheel, on the winding (spring) arbor transmits power from the spring to the pinion of the second arbor. This is a solid pinion. All other pinions are of the lantern type.

c. The count wheel is also on the first arbor, but is driven by a solid pinion on the second arbor, rather than by a pin on the second wheel, as was the case with the 30 hour movement. It is secured by a wishbone friction clutch washer.

d. The second wheel is mounted near the center of the second arbor and engages with the pinion of the third arbor.

e. The third wheel is mounted near the rear end of the third arbor and engages

with the pinion of the fourth arbor.

f. The strike locking cam is also mounted on the third arbor. This circular cam, with a rectangular notch and a flat section, has two pins attached to it. These pins lift and drop the strike hammer lever.

g. The fourth wheel is mounted on the fourth arbor and engages with the cam of the fly arbor. The warning stop pin is mounted on the back side of the rim of this wheel.

h. On the fifth arbor is the fly governor, which is made in one piece, of sheet spring brass. At its middle, it has an integral friction spring section with a dimple to position it in a groove in the arbor.

2. The Time Train:

The time train in this case involves only transmission of power to drive the escape wheel and, through it, to pendulum. To reduce the rate at which power is released from the spring, so the movement can run for eight days, it has two more

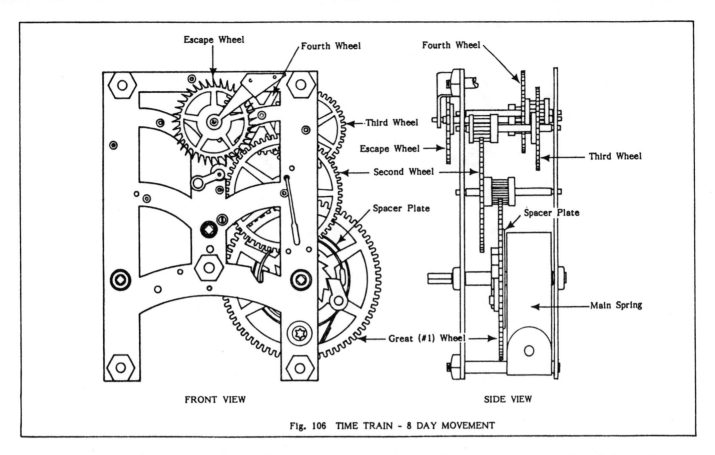

FRONT VIEW   SIDE VIEW

Fig. 106 TIME TRAIN - 8 DAY MOVEMENT

arbors, wheels and pinions than the otherwise similar 30 hour movement discussed in Chapter 10.

a. The main spring of the time train of this movement is much wider than that for the strike train. A thin protective disc, or spacer plate, is mounted between the side of the main spring and the first wheel. This is, presumably, to prevent coils of the spring from being caught on the end of the click spring rivet.

b. The first (great) wheel is mounted freely on the winding arbor, it remains stationary while the arbor is turned, during winding. When winding ceases, the click mounted on a spoke of this wheel engages a tooth of the ratchet wheel, which is securely attached to the arbor. Force is then transmitted from the spring to the arbor, ratchet, click and wheel. This wheel then transmits power to the pinion of the second arbor.

c. The second wheel is mounted on the second arbor and transmits power to the pinion of the third arbor.

d. The third wheel, mounted on the third

arbor, transmits power to the pinion of the fourth arbor.

e. The fourth wheel, mounted on the fourth arbor, transmits power to the pinion of the fifth, or escape wheel arbor.

f. The escape wheel on the fifth arbor, like the one on the 30 hour movement described in Chapter 10, is on the outside of the front plate. The pinion on this arbor must be disengaged from the the fourth wheel before the front plate can be lifted clear.

3. The Motion Work Train:

This train is almost identical to that of the 30 hour movement discussed in Chapter 10. It differs in the location of the transfer arbor, which is above the minute arbor in this case. It receives power from the time train and provides one revolution of the minute arbor in an hour and one revolution of the cannon (hour) tube in 12 hours.

a. The minute arbor carries the strike lifting pin and has a solid pinion at the

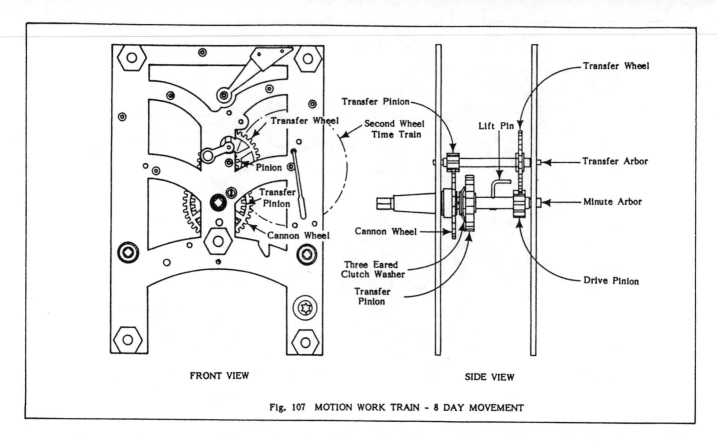

FRONT VIEW   SIDE VIEW

Fig. 107 MOTION WORK TRAIN - 8 DAY MOVEMENT

back end. The cannon tube and its wheel fit over the front end of the arbor.

b. The transfer pinion fits on the minute arbor, but is not attached to it. A three eared spring clutch washer between the cannon wheel and the transfer pinion forces the pinion against a flange on the arbor. The transfer pinion engages with the second wheel of the time train, which is the source of power for the motion work. The clutch washer makes it possible for the arbor to be turned, during hand setting, while the transfer pinion remains stationary.

c. The transfer arbor of the motion work train is found slightly above and to the right of the minute arbor. It has a pinion at the front and a transfer wheel at the back. The wheel engages with the pinion on the minute arbor, transferring power through its pinion to the cannon wheel.

4. The Strike Lever System:

Here, again, the functioning of this system is essentially the same as the one in the 30 hour movement. Some minor differences do exist, however. For

example, in the 30 Hour movement we have described, (Fig. 104) the warning stop lever and the warning lift lever straddle the arbor carrying the other levers. In this 8 Day movement, both of these levers are below the arbor carrying the other levers. The arrangement of the cam and pins is different, because of the additional wheels. Warning locking is done by a pin on a wheel, rather than by a pin on the fly arbor.

These and other minor variations make it important to sketch such details of a particular movement, before disassembly. The relocation of levers and lever arbors is simple if you know how they are supposed to be located, but can be most confusing if you don't.

5. Operation of the Strike Lever System:

This train operates in similar fashion to the 30 Hour one previously described.

a. The lift pin is located on the minute arbor, lifting and dropping the warning lift lever once each hour.

b. The warning lift lever, as it rises, moves the warning lock lever into position

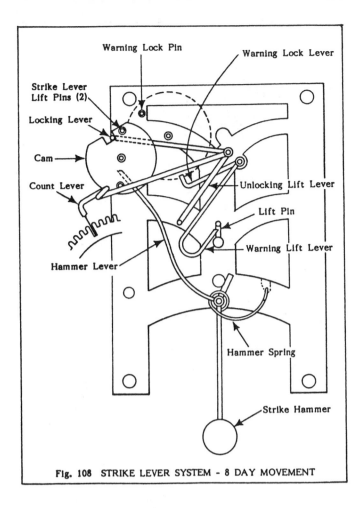

Fig. 108 STRIKE LEVER SYSTEM - 8 DAY MOVEMENT

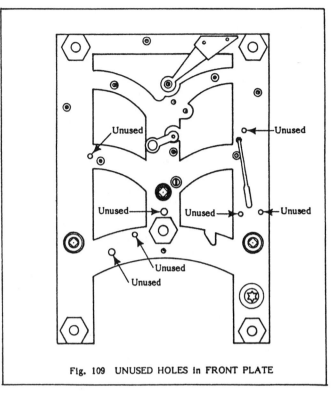

Fig. 109 UNUSED HOLES in FRONT PLATE

to engage the warning lock pin on the fourth wheel. It also engages the unlocking lift lever raising the count lever clear of the count wheel teeth, then lifting the locking lever clear of the slot in the cam, allowing the strike train to move.

c. As the train is released, the warning lock pin hits the warning lock lever and the train is again stopped, until the warning lift lever drops off the lift pin, dropping the warning lock lever and freeing the train to begin the striking sequence.

d. The strike hammer is lifted and dropped by two pins mounted on the locking cam. The strike hammer spring is secured in a notch in the front plate and the hammer stop lever rests on the fifth pillar in the center of the movement.

6. Front Plate:

The front plate is secured to the pillars by nuts, rather than taper pins.

To avoid confusion during reassembly, it is most important to note that there are a number of holes in the plate, which look like pivot holes, but are not. It is probable they were used to position the plate at some stage of the manufacturing process. Some of them are near pivot holes and could be confusing during reassembly. True pivot holes in this plate have oil sinks, depressions around the hole, while the unused holes do not. In other movements, there may be no oil sinks and the problem is compounded.

Carefully examine the front plate of the movement you will disassemble for similar anomalies and make a sketch or written note of them.

CAUTION:

Before starting disassembly, carefully read the following section discussing springs. Springs of your movement, especially main springs, MUST BE IMMOBILIZED before beginning disassembly.

II. MAIN SPRINGS:

In this section, we will discuss various types of main springs, immobilizing, releasing, cleaning, lubricating and reinstalling.

Because of the importance of this subject, we strongly recommend that you read and understand this section in its entirety, before beginning disassembly of your movement. It will not take long and will give you valuable insight into clock main springs and how to deal with them safely.

A. IMMOBILIZING MAIN SPRINGS:

Remember that, unless they are fully relaxed, main springs store energy. This energy can be substantial and must be controlled and contained if damage or injury are to be avoided during disassembly.

The energy contained in a main spring is controlled, in its normal condition in the movement. Energy is transferred to the wheels and pinions of the trains and slowly released as needed.

Before disassembly, the energy of main springs must be disconnected from the train and otherwise controlled. This is done by confining the outer coil of the partially wound spring.

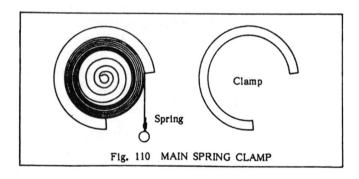

Fig. 110 MAIN SPRING CLAMP

1. Main Spring Clamp:

The main spring clamp shown in Fig. 110 is of a type commonly used to restrain new replacement springs. It is made of round steel in a C shape. The opening permits the end of the spring to project outward so that it can be hooked on the pillar of the movement.

Before letting down a spring, it is wound until the clamp can be fit over it. The clamp will then confine the spring as it is unwound.

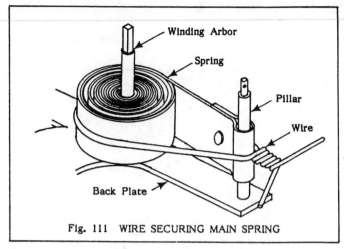

Fig. 111 WIRE SECURING MAIN SPRING

2. Wire Restraint:

Fig. 111 shows how a main spring can be restrained, using soft heavy iron or steel wire. The spring is almost fully wound, then the wire is looped around it and twisted at the loop end secured to a pillar of the movement. When the spring is let down, the wire confines it.

We use soft steel wire, about #18 gauge, which is readily available at most hardware stores.

a. Wind the spring until the outer coil lies about 1/4" inside the outer edge of the main wheel.

b. Cut a length of wire about a foot long and curve one end so you can feed it over the top of the spring and down well past the bottom of the movement. Bring the ends together, just below the pillar on which the loop end of the spring is secured, and twist them, using a pair of pliers, until the loop is in firm contact with the outside coil of the spring.

B. LETTING DOWN the MAIN SPRING:

To release the energy stored in the spring, we must allow the arbor, rather than the wheel to turn. The arbor must be kept from turning, while the restraining click is moved clear of the ratchet. Then the arbor allowed to turn slowly, under control, until the spring is fully relaxed. This procedure is known as 'letting down' the main spring.

For this operation you must have a special tool which will closely fit the square of the winding arbor and a smooth handle which can be gripped in the hand. Release is accomplished by slowly relaxing the grip, so that the handle slips in the palm of the hand, at a controlled rate. There are three basic types of let down tools, available at nominal cost.

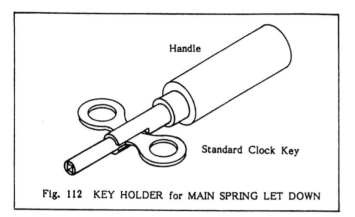

Fig. 112 KEY HOLDER for MAIN SPRING LET DOWN

1. Let Down Key Holder:

This is a simple device, consisting of a wooden handle into which a brass or steel socket piece is inserted. This socket is drilled to receive the stem of a clock key and has a slot for the ears of the key. The advantage of this type is that it uses the same key designed to wind the clock. The disadvantage is that not all keys have the same body diameter, so the fit in the socket may be rather sloppy. This can complicate letting down.

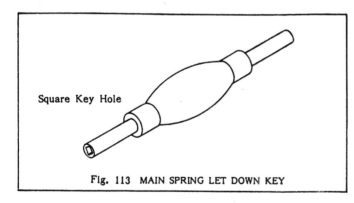

Fig. 113 MAIN SPRING LET DOWN KEY

2. Double Ended Let Down Key:

This type has a socket of different size at each end, that fit the square on the winding arbor. The handle is shaped to comfortably fit the hand, providing good control during the let down procedure. A disadvantage is that only two sizes of key socket are available in one tool.

3. Let Down Key Set:

This is a series of six double ended steel key sockets, providing twelve sizes. The body of each key is hexagonal and fits snugly into a smooth plastic handle. The primary advantage of such a set is that it provides a key of the proper size to fit any winding arbor.

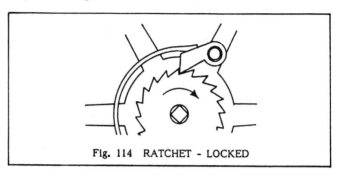

Fig. 114 RATCHET - LOCKED

4. Procedure for Letting Down Main Springs:

a. Click and Ratchet Locked:

Main springs are hooked to the winding arbor and wrap around it during winding. The ratchet wheel turns during winding, while the main wheel remains stationary. The click, mounted on the main wheel, is repeatedly lifted and dropped between ratchet teeth as winding occurs.

When winding is complete, the click locks the ratchet and the wheel together, so that power is transmitted from the ratchet to the wheel. The position of the click engaged with the ratchet is shown in Fig. 114, as they will be when you start.

b. Install Spring Restraint:

Before proceeding, wind the main spring until you can place the main spring clamp around it. If you use the wire method, install it as described above.

c. Disengaging the Click:

This procedure should be carefully followed when you are getting ready to let a main spring down. The click and the tool used to move it must be kept clear of the teeth of the ratchet wheel during unwinding. The let down key must be held firmly and allowed to turn only slowly.

If the winding arbor is allowed to turn rapidly and the click is accidentally released, serious damage to the click and ratchet teeth can result.

(1) Position the Click:

The click is seldom found outside the plate, as shown in Fig. 106, when you want to release it. With the type of movement we show, it is a simple matter to get it into this position, where it is accessible.

For the time train main spring, prevent the escape wheel from moving by placing a finger on it, then remove the anchor. Slowly release finger pressure and allow the escape wheel to turn, until the tip of the click moves well outside the plate.

The strike train main spring can be treated in similar fashion by lifting the count lever just above the teeth of the count wheel and holding it there. The train will run without stopping. Continue until the click moves outside the plate.

(2) Engage Let Down Key:

Place the let down key on the winding arbor, making sure it is of the proper size to fit snugly and is fully engaged. Grip it firmly in one hand.

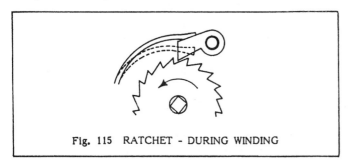

Fig. 115   RATCHET - DURING WINDING

(3) Disengage Click:

Click disengagement can be somewhat simplified if the click spring is of the round type, with a flattened end fitting into a slot in the click, as shown in Fig. 115. Before starting to wind the arbor or lift the click, move the spring tip out of its slot and let it come to rest at the side of the click. If the clicks of your movement are not of this type, you must

overcome the force of the click spring during disengagement.

With a small screw driver or other small pointed tool in the second hand, turn the winding arbor slightly to bring the point of the click just clear of the bottom of the click tooth, as in Fig. 115.

Hold the let down tool firmly and, with the tip of the screw driver under the click, lift and hold the click well above the tips of the ratchet teeth. You must continue to hold it securely in this position during let down.

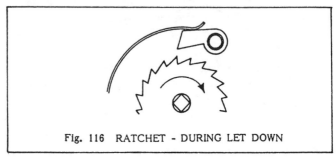

Fig. 116   RATCHET - DURING LET DOWN

c.   Begin Spring Let Down:

Make sure the restraining clamp or wire is in position around the spring.

While holding the click clear of the ratchet, gradually release your hand pressure on the let down key and allow it to turn smoothly. Continue, until you feel no further power from the spring.

When both main springs have been let down, you may proceed with disassembly. See II of Chapter 10.

B.   TYPES of MAIN SPRINGS:

There are two basic types of main springs. One is open and the other is confined within an enclosure called a 'barrel'.

1.   Open (Loop End) Main Springs:

All main springs are attached to the winding arbor by means of a hole in the end of the spring which attaches to a pin on the arbor. Loop end main springs are usually unconfined. That is they are free to expand as they unwind, until the they are completely unwound, or the outside

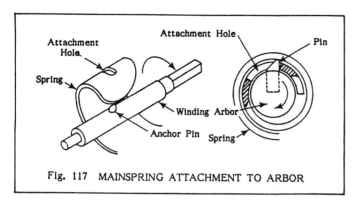

Fig. 117 MAINSPRING ATTACHMENT TO ARBOR

coil meets some obstruction.

These springs get their name from a loop formed at the outer end, which fits over a pillar of the movement. This fixes the outer end of the spring.

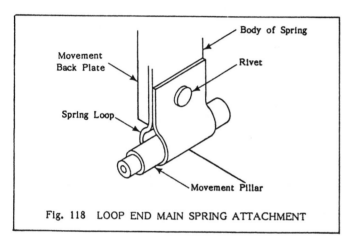

Fig. 118 LOOP END MAIN SPRING ATTACHMENT

a. Broken Loop End Main Springs:

If this type of spring is broken, it has essentially uncoiled and lost most of its energy. It will probably be at least partially confined by elements of the movement, or the case. During disassembly of the movement, there will be some disruptive action when the plate is removed and the remaining spring force is released.

There is no certain way of eliminating this problem. Usually, the greatest difficulty will occur in taking the movement out of the case, if the outer coils of the spring are in contact with the case. When they are freed from the case, the spring will further unwind and most of its energy will be dissipated.

Unbroken open main springs can cause a great deal of trouble if disassembly of

the movement is attempted before they are confined and their energy released, under control, until restrained by the clamp or wire. We realize this is a repetition of earlier warnings, but it may save you serious injury.

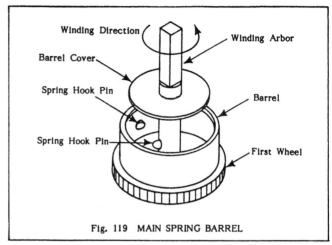

Fig. 119 MAIN SPRING BARREL

2. Main Springs in Barrels:

A barrel is a ring of brass, slightly wider than the main spring, which is attached to the first wheel. A close fitting cover fits in a groove at the end opposite the wheel. It effectively limits the outward expansion of the coils of the spring as it unwinds.

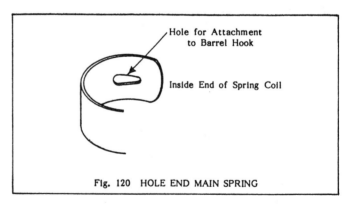

Fig. 120 HOLE END MAIN SPRING

Unlike loop end main springs, those designed for use in a barrel have a hole, rather than a loop, at their outer end. This hole engages with a short pin on the inside of the barrel. For this reason, they are referred to as 'hole end springs'.

While springs in barrels are contained, they always store considerable amounts of energy, even when completely unwound, or even broken, but confined within the barrel. This is because the barrel prevents full relaxation of the coils of the spring.

Before starting movement disassembly, the spring must be allowed to fully unwind within the barrel, so that no force is applied to the winding arbor. If the spring is broken, or unhooked, there will be no resistance when winding is attempted and the following procedure will not be required.

a. Letting Down the Main Spring in a Barrel:

We have learned that a ratchet and click wheel allow the winding arbor to be turned to tighten the spring around the arbor, then lock in order to transmit stored power to the wheel train.

Since the barrel will effectively restrain the spring when it is completely unwound, not other restraining device is needed.

In order to safely release this energy, place the let down key on the winding arbor, then turn it only enough so that the click is not bearing against a tooth of the ratchet wheel. Holding it there, use a small screw driver to lift the click clear of the teeth of the ratchet, then ease the pressure on the handle of the let down key so that it rotates slowly in your hand, until all the force of the spring is dissipated.

The coils of the spring are now safely nested against the wall of the barrel, which acts as a retainer.

When both main springs are fully let down, you can safely proceed with disassembly, following the procedure beginning at II of Chapter 10.

C. FULLY RELEASING MAIN SPRINGS
    (After disassembly of the movement.)

CAUTION: Main Springs should be released completely only when they are not functioning properly and cleaning and lubricating is necessary to restore their usefulness

For cleaning, mainsprings must be completely unwound and relaxed. While the preferred method of doing this is to use a comparatively expensive clock maker's spring winding tool, it can also be accomplished with a much less expensive tool, or without any such equipment.

1. Loop End, Wire Tied Main Springs:

a. Relieving Spring Force:

Having confined your spring with wire, it is now a simple matter to release it.

CAUTION: Remember that a lot of force is stored in the spring. Follow these instructions carefully.

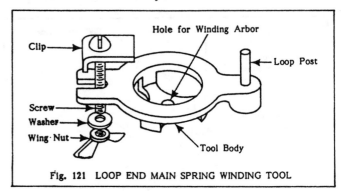

Fig. 121 LOOP END MAIN SPRING WINDING TOOL

(1) Simple Loop End Spring Winding Tool:

A simple loop end spring winding tool, suitable for use with arbors 3/16" or less in diameter, is available from clock parts supply houses at a cost of less than $10.00. It consists of a casting with a recessed section, in the center of which is a hole to receive the winding arbor. A clip with a screw and nut for securing the wheel of that arbor is at one end and a pin on which the loop end of the spring is fitted is at the other.

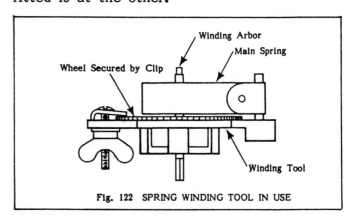

Fig. 122 SPRING WINDING TOOL IN USE

After it has been taken out of the movement, take the assembly of mainspring, arbor and wheel and insert the arbor in the hole of the tool, with the spring uppermost. Make sure the click is free in the opening in the tool. Fit the tool clip over the rim of the wheel and tighten the wing nut. Using the let down key, wind the spring until the wire loop is loose, then lift the loop off.

Hold the part of the tool on which the pin is located firmly in a vise. Using a let down key, wind the spring just enough to free the click from the tooth against which it bears. With the point of a small screwdriver, move the click clear of the ratchet. Slightly release the pressure on the let down tool and slowly allow the spring to fully uncoil.

Unhook the spring from the arbor and clean the arbor and wheel assembly, along with other parts. Cleaning an lubrication of the spring will be discussed later.

(2) Without a Spring Winding Tool:

Secure a bath towel, or similar large bulky piece of material. WEAR A GLOVE. Position wire cutters on the open length of the wire loop that lies between the body of the spring and the spring loop. DON'T CUT YET. Wrap the towel around the spring and then cut the wire. The spring will uncoil violently, but the towel will keep it under control. Unwrap the towel.

Unhook the spring from the arbor and clean the arbor and wheel assembly, along with other parts, before reassembly. Cleaning and lubricating the main springs will be discussed later.

2. Barrel Retained Main Springs:

a. Remove Barrel Cover: (Fig. 119)

A brass plate covers the spring at the open end of the barrel. It fits snugly into a groove on the inside of the barrel. To remove the cover, grasp the assembly firmly in one hand, with the cover side down, then tap the end of the arbor sharply on a hard surface. The inertia applied to the spring applies force to the

cover. Repeat until the cover comes loose and can be removed. If this is not successful, after several taps, turn so that the cover side is up and tap the opposite end of the arbor on a hard surface.

b. Relieving Spring Force:

Without special equipment, the spring must simply be pulled from the barrel. This procedure may cause some minor distortion of the spring and breakage may occur.

BEFORE PROCEEDING, it is suggested that you read instructions for reinserting the spring in the barrel, given at D.2, below.

CAUTION: Significant energy will be instantly released when the spring comes clear of the barrel. The spring will not be restrained by a wire tie or clip. Follow instructions carefully.

Broken Springs in barrels still retain a great deal of force, much like those that are intact, since they are still confined within the barrel. They also have sharp ends, which can cause serious injury.

It is important to protect your hands with heavy leather gloves.

Work in a standing position, with plenty of room around you.

Use long nose pliers to grip the center coil of the spring. With the other hand, wrap the towel around it and hold the barrel in a firm grip, then pull the spring from the barrel. Use as much force as necessary.

The spring will jump out of the barrel rather violently. Be prepared.

Clean the barrel assembly, along with other parts, before reassembly.

C. CLEANING and LUBRICATING
     MAIN SPRINGS:

This procedure is applicable to both types of main springs.

Now that the coils of the spring are much more widely separated, you should be able to see areas of rust and corrosion. Since, in use, the faces of the spring bear against each other with considerable force, surfaces that are not completely smooth act as brakes and may stick so tightly that movement of any kind is impossible. This is a common reason for stoppage of a movement.

Our first objective is to clean and smooth the entire length of the spring on both sides.

1. Cleaning:

Soak a small pad of cloth in your clock cleaning solution and, covering both sides of the spring, scrub back and forth entire length. When you come to the area where the coils are very close together, try using two popsicle sticks with a fold of cloth around them, to reach a bit deeper into the coils.

Change the cloth surfaces frequently, as they become dirty.

When the spring is clean, so that no dirt shows on a new cloth surface, thoroughly rinse the entire spring in hot water. Carefully dry the spring as quickly as possible. Use a hair dryer for final drying, if you have one.

2. Polishing:

Use very fine steel wool, or #400 wet-or-dry sand paper to polish out all rust or other rough spots. It will be easier, particularly if there are heavy rust areas, if you work against a firm surface.

Where the coils are close together, wrap a popsicle stick with sandpaper or steel wool and clean as far as you can. Carefully remove dust and bits of steel.

Continue polishing until the entire surface is smooth to the touch. Wipe all surfaces carefully, to remove any residue from the steel wool or sand paper

3. Lubricating:

Only a film of lubricant is required on the surfaces of the spring. Avoid over lubricating.

Use only oil or grease specifically designed for clocks, which is available from clock parts suppliers. DO NOT USE HOUSEHOLD OIL, MOTOR OIL OR SPRAY LUBRICANTS.

Apply the lubricant sparingly to a soft cloth pad and wipe both sides of the entire length of the spring.

D. RECOILING CLEANED and LUBRICATED SPRINGS:

1. Loop End Springs:

a. Attach Spring to Arbor:

Do this only after the arbor and wheel assembly has been cleaned.

Select the proper main spring arbor, and check the winding direction by holding the wheel and turning the arbor in the only direction permitted by the click and ratchet. Locate the inside loop of the spring over the arbor so that, when hooked to the arbor, it will wind in the proper direction, then locate the hook in the hole at the end of the spring.

b. Using the Spring Winder:

Rewinding a main spring, using this tool, is simply a reversal of the procedures outlined for letting down the spring.

(1) In use, the loop of the mainspring is placed over the pin, the arbor inserted through the hole. The wheel, which lies flat against the face of the tool, is secured in place by tightening the wing nut to clamp the wheel against the face of the tool. The wheel is thus secured and kept from turning, while the spring is secured by the pin.

(2) Using the key, the arbor may now be turned to wind the spring. The ratchet and click will prevent unwinding. Wind until the outer coil of the spring is just inside the outer rim of the wheel.

(3) Now, as you did before disassembling

the movement, place a steel rod main spring clamp around the spring, or replace the wire loop you formed to restrain the spring before disassembling the movement.

(4) Using the let down key, turn the arbor just enough to permit release of the click with a small screw driver and allow the arbor to turn until all the force of the spring is held by the clamp or wire and none is exerted on the arbor. Reset the click so that it bottoms between teeth of the ratchet wheel.

(5) Loosen the clamp screw and remove the arbor from the winding tool. It is ready for reassembly into the movement.

c. Recoiling Without a Loop End Spring Winder:

Main springs can be recoiled without the aid of a winding device, but extra caution is required.

(1) Select the proper arbor for each spring and check for proper winding direction, then insert the arbor into the center coil of the main spring and make sure the hook on the arbor is properly engaged in the hole of the spring.

(2) Place the loop of each spring over the proper movement pillar and insert the pivot of the arbor in its hole in the back plate.

(3) Take the front plate and insert the end of each mainspring arbor in its proper hole, then lower the plate onto the pillars. Screw nuts onto each of the pillars and tighten them.

(4) Wind each spring until a main spring clamp can be slipped over it, or, if you are going to use wire to secure it, until the outer coil is about 1/4" inside the outer rim of the wheel and secure with wire, as you did before disassembly.

(5) Release the click and, using the let down tool, allow each spring to uncoil until restrained by the clamp or wire.

(6) Again remove the front plate and proceed with assembly of the movement.

2. Recoiling Springs in Barrels:

Be sure the barrel assemblies have been cleaned.

Particularly with heavier springs, this is a difficult and possibly dangerous task. It may result in the spring breaking. Wear leather gloves and eye protection.

a. Verify the proper direction of rotation. The pin on the inside of the barrel usually has a tapered side and a flat side. The flat side of the pin will be in contact with inside of the hole in the spring nearest the end of the spring.

Angle the end of the spring and hook the hole over the pin. Make sure the hole in the spring is firmly secured on the pin and remains so as you continue.

b. Grasp the barrel firmly with one hand and begin feeding the spring into the barrel, a little at a time. It will be necessary to hold the coils in the barrel with the thumb of the hand gripping the barrel. Continue until the spring is entirely inside the barrel.

c. Insert the winding arbor into the inner coil of the spring and, by turning, seat the pin on the arbor in the hole near the inside end of the spring. Be sure the straight side of the pin bears against the side of the hole nearest the end of the spring. Test with the key to make sure it is firmly engaged. It may be necessary, using long nosed pliers, to bend the end of the spring so that it engages the pin securely.

d. Slip the barrel cover on the arbor, making sure the finished face is up. Locate the edge of the barrel on a firm surface. Place a small piece of wood or heavy cardboard against the face of the barrel cover and, using your small hammer, tap lightly near the edge of the cover, turning the barrel and tapping until the cover is seated.

The barrel assembly is now ready to be reassembled into the movement. See procedure for reassembly, Chapter 10, IV, substituting the spring barrel arbors for the winding drum arbors of the weight driven movement and adding the fourth and fifth arbors in each train.

E. REPLACING BROKEN MAIN SPRINGS:

If the clock is to operate properly, the power supplied by new springs must, as closely as possible, duplicate that of the originals. Most springs can be replaced with identical duplicates. Clock parts supply house catalogs list a very wide variety of springs with minute differences in important dimensions, particularly thickness.

If you want to replace more than one spring in a movement, carefully measure each one of them. Dimensions of springs in the same movement can be very different, but even minor differences are important.

1. Spring Type:

You must identify your spring as either LOOP end (unenclosed) or HOLE end (barrel enclosed).

2. Measuring the Spring to be Replaced:

a. Width:

It is important that width be measured to the nearest 1/64" or 0.1 (1/10) millimeter. Use your scale.

b. Thickness:

This must be measured with a micrometer. If you do not have one, you may be able to have a friend measure it for you. The measurement should be accurate to the nearest .001" or 0.01 millimeter.

c. Length:

Measurement of length to the nearest inch will usually be adequate. Because a main spring cannot be fully straightened out, measurement is difficult. You may find a cloth tape, such as used in sewing, helpful, when you come to the inner coils.

3. How New Springs are Received:

Main springs are partially wound and restrained at the factory and are usually ready for installation in the movement.

a. Wire Tied:

When springs are shipped with a wire tie, they are restrained in much the same manner as we have discussed. See Fig. 111

b. Clamp Ring Retained:

In this method, a ring of heavy steel rod, with a gap in it, is used to restrain the spring. See Fig. 110.

F. INSTALLING NEW SPRINGS:

1. Loop End:

Check for correct direction of winding, then insert the winding arbor into the spring and make sure the hole in the spring is securely attached to the pin on the arbor.

In some cases, wire tied loop end springs will have enough room to permit inserting the arbor in the plate, after the loop has been placed over the pillar of the movement. If this is not the case, it will be necessary to follow the procedure outlined above to release it and rewind it on the spring winder, or between the plates of the movement.

If the spring is retained by a clamp, the tail with the loop in it usually sticks out far enough to permit assembly. If it does not, put the spring on the winding arbor and assemble on the winding tool, then wind it enough to reset the ring, so that the maximum amount of the tail is free.

2. Hole End (Barrel Enclosed):

These springs are usually retained by a clamp ring, with the hole end projecting out from the gap in the ring.

If you know someone who has a spring winding machine, it would be wise to ask him to insert the spring in the barrel for you.

a. If you must do it yourself, locate the pin on the inside of the barrel. When you insert the end of the spring, make sure the straight side of the pin on the inside of the barrel will be at the side of the hole nearest the end of the spring.

b. There should be a fair amount of clearance between the outer coil of the spring and the inside of the barrel. Angle the spring so that you can hook the hole in the spring over the pin.

While turning the spring to keep it engaged with the pin, lower it into the barrel until the steel retaining ring bears on the edge of the barrel. This will require some force.

c. The retaining clamp will now bear on the upper edge of the barrel. Place a small block of wood on the face of the coiled spring and gently tap with your small hammer, to push the spring out of the ring.

As you near the point where the clamp ring is about to clear the edge of the spring, check to make sure the spring is hooked on the pin. As you continue tapping, the ring will come free and the spring will expand outward against the inside of the barrel.

d. Insert the arbor, again making sure the pin on the arbor is properly engaged in the hole in the spring. Using the key, test turn the arbor to make sure.

e. Replace the cover and seat it in its groove, using a small block of wood and your small hammer.

The barrel assembly is now ready for reassembly into the movement.

III. CLEANING THE MOVEMENT PARTS:

Follow the procedures in III of Chapter 10. If yours is a 30-hour spring driven movement, the only basic difference will be that the first arbor will have a spring instead of a winding drum.

IV. REASSEMBLING an 8 DAY SPRING MOVEMENT:

A. PROCEDURE:

Since 8 Day movements have more wheels and arbors than 30 Hour movements, a slightly different procedure makes reassembly easier, particularly with respect to proper location of the levers of the strike system. This involves installation of the arbors of the strike train in the back plate, before locating those of the time train.

1. Install assembly clamps on the back plate.

2. Place the loops of both main springs over their respective pillars. As a check that you have the springs in the right location, note that usually the loop side of the spring will be toward the outside of the movement.

3. Insert the pivots of both main spring arbors in the back plate. This may require a little force to overcome resistance of the spring, to move the pivot into position.

4. Place the cannon tube and wheel on the minute arbor of the time train and insert the pivot of the minute arbor in its hole near the middle of the back plate.

5. Insert the motion work arbor in its hole in the back plate, near the minute arbor, so that its wheel and pinion engage with those of the minute arbor and cannon wheel.

6. Insert the pivot of the second wheel of the strike train in its hole in the back plate, making sure that its pinion engages with the teeth of the first wheel.

7. Insert the pivot of the arbor with the lift and warning levers in its hole near the top of the back plate. The outside of the hook of the lift lever must rest on top of the minute arbor.

8. Insert the pivot of the arbor with the count and lock levers in its hole in the back plate, usually just above the lift lever arbor.

9. Insert the pivot of the third wheel, with its cam, in the back plate, making sure its pinion engages with the teeth of the second wheel.

a. Seat the lock lever so that it is in the bottom of the slot in the cam.

b. The strike count lift lever should rest on top of the cam, between the strike lever activating pins that project from the side of the cam.

10. Insert the pivot of the fourth wheel in its hole in the back plate, making sure its pinion engages with the teeth of the third wheel.

11. Insert the pivot of the fly arbor in its hole in the back plate, making sure its pinion engages with the teeth of the fourth wheel.

12. Insert the pivot of the second wheel in its hole in the back plate, making sure its pinion engages with the teeth of the first wheel.

13. Insert the pivot of the third wheel in its hole in the back plate, making sure its pinion engages with the teeth of the second wheel.

14. Insert the pivot of the fourth wheel in its hole in the back plate, making sure its pinion engages with the teeth of the second wheel.

15. Escape Wheel Arbor:

a. Escape Wheel in Front of Front Plate:

If your movement has an outside escapement, with the wheel and crutch operating in front of the front plate, the end of the escape wheel arbor on which the pinion is mounted should now be run through the opening in the front plate near the bracket containing the escape wheel pivot hole. The wheel will then rest on the face of the front plate.

b. Escape Wheel Inside Front Plate:

If the escapement of your movement is between the front and back plates, insert the pivot of the escape wheel arbor in its

hole in the back plate, making sure the pinion engages with the teeth of the fourth wheel.

The anchor and crutch arbor should then be positioned in the back plate.

16. Lower the front plate of the movement over the minute arbor, then over the two winding arbors.

17. Position the hammer arbor pivot in its hole near the bottom of the back plate, making sure the activating tail is positioned in line with and above the pins on the strike train wheel.

18. Carefully align holes at the bottom of the front plate with the two bottom pillars attached to back plate, and lower it over them. Place a nut on each post and turn about one turn to loosely secure the plate.

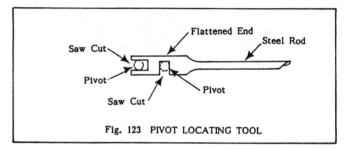

Fig. 123 PIVOT LOCATING TOOL

You will find a pivot locating tool, like the one shown in Fig. 123, very helpful in locating pivots in the front plate. Ready made ones can be purchased from clock parts suppliers, or you may want to make one.

A piece of coat hanger wire or the #18 wire suggested for securing main springs does very well.

Cut a piece of wire about 8" long and flatten one end, by pounding it lightly with a hammer on a flat solid steel surface. File its sides and end, then cut a small groove in the end and one in the side, as shown. Hold the piece in a vise and gently use a hacksaw to cut the grooves. The grooves should be just a little wider that the diameter of the largest arbor. Smooth the edges before using.

In use, the groove on the end is used

to push a pivot toward its hole, while the one in the side is used for pulling.

19. Working from the bottom of the movement toward the top, while applying slight finger pressure to the upper end of the top plate, locate the upper ends of successive pivots of the time train in their holes in the front plate.

Use the pivot locating tool for positioning the pivots. If you don't have such a tool, a pair of tweezers may be helpful. Providing only little force is used, a surgical clamp called a hemostat or forceps is very good for this operation.

20. If your movement has an escapement outside the plates, locate the bottom pivot of the escape wheel arbor in its hole in the back plate.

21. Loosely attach a nut to the top pillar on the time train side.

22. Continue by locating pivots of the strike train in their holes in the front plate. Make sure that levers are in proper position, particularly that the count lever is in a deep slot of the count wheel and the locking lever in the cam slot.

23. Holding the top plate down gently, with a finger tip, move each wheel up and down. If the pivot is in the hole, there should be some movement. If a pivot is not in its hole, there will be no movement.

Check for proper location in both top and bottom plates. Using light finger pressure on the teeth of the second wheel in each train, turn it back and forth slightly to check for free movement of all the wheels.

24. Put nuts on the remaining pillars and tighten all nuts. Remember that these are brass nuts, easily stripped if too much pressure is applied, so use only enough to make them snug.

25. Rehook the ends of the coil springs on the three lever arbors over the edge of the movement.

26. If you have an outside escapement, locate the holes in the anchor pivot bracket on their pin and secure by rotating the spring retainer so that it rests on the end of the pin.

27. Lightly lubricate each pivot, using CLOCK OIL ONLY. As described earlier, this may be done by dipping a toothpick in the oil and touching it to the outside of the pivot at each hole.

You may be ready to acquire a special clock oiler, a tube with a fine needle-like tip. These are available from clock parts suppliers, sometimes filled with clock oil. When using an oiler, use very special care to apply only a tiny amount of oil to each pivot.

28. Also, using the same method, lightly oil the face of each anchor pallet where it comes in contact with teeth of the escape wheel.

IV TESTING:

If you have carefully followed the above procedure, the movement should now operate satisfactorily, unless there were defects left uncorrected.

Wind the time side main spring only until the outer coil is free from the retaining wire or clamp. With the movement in its normal vertical position, move the crutch back and forth and observe the action of the teeth of the escape wheel as they strike the pallets of the anchor. The teeth should strike the pallets sharply, indicating that the movement is free and efficiently transmitting power.

Wind the strike side main spring only until the outer coil is free from the retaining wire or ring. Place the minute hand on its arbor and slowly turn clockwise. Watch the count lever, which, if levers and wheels have been properly positioned, should be lifted out of its slot and the strike train activated slightly until the warning lever stops it. Continue turning until the count lever drops and striking begins.

If the strike train does not function

properly, you have probably not located wheels and levers correctly. Recheck the location of all elements in your strike train and lever system against the illustrations in this Chapter and in Chapter 9. Examine the entire train carefully and locate any discrepancies.

Let down both mainsprings against their clamps or retaining wires, then disassemble the movement and make necessary corrections. Reassemble the movement. Don't be discouraged if this happens to you. Nearly all of us have made a goof that made it necessary to start over.

When both trains are functioning properly, FULLY WIND BOTH SPRINGS AND REMOVE RETAINING RING OR WIRE.

Remount the movement in its case, install suspension spring and pendulum and proceed to make any necessary timing adjustments as described in Chapter 5 and Chapter 10.

Chapter 12

BASIC REPAIRS AFTER DISASSEMBLY

Introduction

In previous Chapters we have outlined minor repairs and adjustments that can easily be made by the novice, without disassembly of the movement. Here we will go into somewhat more advanced techniques which, in some cases, will require modest investment in special hand tools, but no major specialized equipment.

The repairs we will discuss in this Chapter are among those most commonly necessary to restore a movement to operating condition and can be done successfully by most people with average mechanical skills. In each instance, disassembly of the movement, as covered in Chapters 10 and 11, must be accomplished, so that parts are accessible for repair.

I. CLICKS and CLICK SPRINGS:

In Chapter 8, beginning at I-D, with Figures 54, 55 and 56, we discussed clicks and click springs in general and some of the problems related to them.

The most common problem with most clicks is that the pivot pin, on which the click moves, becomes loose and no longer holds the click securely in its proper position. The pivot pin is secured to the main wheel by riveting, or peening over, its end. All too frequently, the manufacturer used pins that were too short and left only very little material for riveting, or did not rivet tightly.

With time and the stress applied to the click and pivot pin by the force of

the main spring, or weight, the rivet progressively loosened and finally reaches a point where the click becomes so loose it will not properly engage the teeth of the ratchet. When this condition approaches that shown in Fig. 55, both the click and the pivot pin should be replaced.

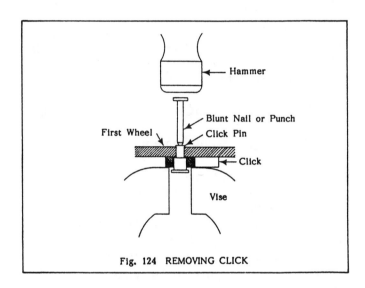

Fig. 124   REMOVING CLICK

A. REMOVING CLICKS:

Clicks are nearly always attached to a spoke (crossing) of the main, or first, wheel. The ratchet is either firmly secured to the winding arbor, or has a square hole which matches the winding square, on which it fits. In the latter case, the ratchet may be removed from the arbor, somewhat simplifying removal of the click.

Removing a click involves forcing the riveted end of the pivot pin back through the hole in the wheel.

1. Procedure:

A punch, smaller in diameter than the hole in which the pivot pin fits, will be used to drive the pin through the wheel. A small nail, with the point flattened by filing, makes a good punch for this purpose. Make sure the diameter of the punch or nail is less than that of the hole in the wheel, so it can push the pin all the way through.

The click and the wheel must be supported in such a way that force can be applied to the riveted end of the pivot pin. A simple way of accomplishing this is with a small vise, with the jaws open just enough to clear the head of the pin, as shown in Fig. 124. Place the wheel assembly so that the head of the pin is centered in the opening between the vise jaws. Make sure that the click is well supported.

You will have to hold the nail and the wheel in one hand, while using a hammer in the other to very lightly tap the nail until the pivot pin is free.

B. REPLACING CLICKS and PIVOT PINS:

1. Obtaining a Replacement Click and Pivot Pin:

A wide variety of clicks is usually available from major clock parts supply houses. While you may not be able to find one identical to the original in your movement, it is usually possible to rather easily modify a click to fit your needs. Some click shapes come in several sizes and are usually sold as an assortment of those sizes, with suitable pivot pins. Available shapes are illustrated in supplier's catalogs and you can select the one nearest to your original.

2. Styles of Clicks:

In Fig. 125 we have illustrated several basic click shapes available at the time of this writing. Variations in detail may occur with each of these shapes, while retaining the same general style. Refer to the letters in the Figure:

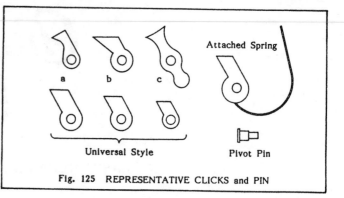

Fig. 125 REPRESENTATIVE CLICKS and PIN

a. This style is commonly used on smaller movements, where it is advantageous to locate the pivot pin close to the ratchet wheel and permit proper engagement with small ratchet teeth.

b. A common style used in many American clocks.

c. Double-ended, or symmetrical. In principle, at least, the symmetrical shape would permit this click to be removed, turned over and reinstalled to provide a new engaging point, using a new pivot pin. It has a small tail for use as a lever handle, making it easier to hold the click clear of the ratchet teeth during let down of the main spring.

d. Attached spring style. Several shapes are available with spring attached, to match originals on many American clocks.

Some come with the spring separate, but with a hole or groove provided in the body of the click, into which the spring must be inserted, then riveted to secure it.

e. Universal style. Here we have shown the three sizes in which this type is supplied. While fitting some movements, as is, this rather bulky type can easily be modified, by filing, to fit a wide variety of conditions. (See Figure 126) The same pivot pin fits all sizes.

3. Fitting Clicks:

In Figure 126, we have illustrated why and how a click must sometimes be modified to obtain proper engagement with the teeth of the ratchet.

Before permanently installing a new

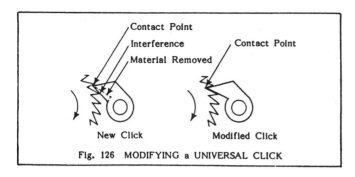

Fig. 126 MODIFYING a UNIVERSAL CLICK

click, it should be placed in position and, with pressure applied to it by a finger tip, the ratchet wheel rotated in the winding direction, then reversed, as if the spring were applying pressure. In the case illustrated, the tip of a tooth prevents the end of the click from bottoming at the base of its tooth, as it should. If used in this condition, it might lock at the tips of the ratchet teeth for a short time, but wear would soon prevent it from engaging even the tip of a tooth. Something must be done.

The new click used is one of the Universal type, as illustrated in Fig. 125. The position of the the ratchet wheel and the hole for the pivot pin are fixed. To make it work, it will be necessary to remove material from the click by filing, as indicated by the dotted line. This new shape permits proper engagement of the point of the click at the base of a ratchet tooth.

a. Filing:

For filing small parts, such as a click, a small file should be used. As you proceed further with clock repair, you will find an inexpensive set of needle files, currently under $10.00, a valuable addition to your list of tools. An alternative is an ignition file, available at most automotive supply outlets.

Hold the click firmly in a vise, positioned so that the face you will file is conveniently located, with the line you want to cut to right at the top of the vise jaws.

Use only a pushing motion of the file. Lift it off the work on the return stroke.

Most files cut on all faces. Carefully position the file before making each cutting stroke and be careful not to cut with the side of the file in an area not to be reduced.

b. Fitting the Click to the Ratchet:

As you progress with shaping the click, remove it from the vise from time to time, insert the pivot pin and locate the assembly temporarily in position. Check your progress in obtaining proper fit. As you approach the final form, check frequently. It is axiomatic that material once removed cannot be replaced.

IMPORTANT: Make sure the tip of the click is at the base of the tooth and that there is a slight clearance between both faces of the click point and the faces of the teeth. Ideally, the shape of the point of the click should closely match that of the space between ratchet teeth.

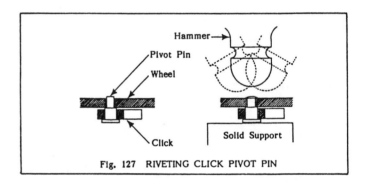

Fig. 127 RIVETING CLICK PIVOT PIN

4. Installing Clicks:

After properly fitting the click to the ratchet, locate the wheel, ratchet and pivot pin so that the head of the pivot pin is on a solid support, preferably a heavy flat steel surface. Make sure the face of the wheel is parallel to the flat surface, so the pin is perpendicular.

Using the ball end of your small ball peen hammer, LIGHTLY tap the end of the pivot pin, varying the angle of stroke as shown. Lift the assembly from time to time and check for tightness and alignment. It should take only a few hammer strokes to securely rivet the pin.

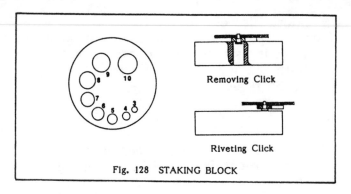

Fig. 128 STAKING BLOCK

5.   Using a Staking Block:

This very useful and inexpensive tool (under $5.00 at this writing) simplifies both click removal and reassembly and has a number of other uses. It is a round steel block, a little less than 2" in diameter with a series of holes ranging from 3 mm to 10 mm in diameter. Since the most common pivot pins have a head 4 mm in diameter, we have shown it using the 5 mm hole for removal.

There is an area of the block free of holes, sufficiently large to be useful for backing up the head of the pivot pin for riveting, during reassembly.

As you progress further with your repair activities, you will find many other uses for this handy tool.

II. CLICK SPRINGS:

We discussed the various basic types of click springs and illustrated them in Figs. 54, 55 and 56.

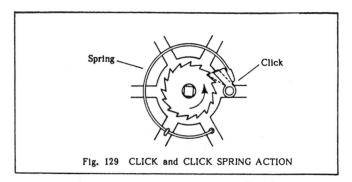

Fig. 129 CLICK and CLICK SPRING ACTION

While your movement is disassembled, it is a good idea to check the action of main wheel click springs. Using the key on the winding square, slowly rotate the ratchet in the winding direction and carefully observe the movement of the click. As it passes the top of a tooth, it should drop smartly to the bottom of the next one.

With the click bottomed and a little pressure on the winding key in the winding direction, use the tip of a small screw driver and gently lift the click. Does the spring exert some pressure at the very bottom? It should. See if the click can be moved, even slightly, without contact with the spring. If it can, the spring is not functioning properly and should be tightened.

In some cases, the spring is so relaxed that it will not push the click between the teeth at all. If this is the case, you most probably noticed, before you started disassembly, that the spring could not be wound, because the click did not lock the ratchet to retain the force put into the spring by winding.

A.   ADJUSTING CLICK SPRINGS:

As we have noted previously, there are several types of click springs. If yours is one that is attached to the click, (Fig. 55), it is a simple matter to unhook it from the clip on the wheel and bend it to a more open position, using the fingers.

CAUTION: Click springs should apply just enough force to the click to make sure its point bottoms between teeth. Excessive pressure will result in wear of both click and ratchet and will cause a loud clicking noise during winding, which is objectionable to many people.

Round or flat springs, permanently attached to the wheel, must be dealt with differently.

1.   Using a Bending Tool to Adjust Fixed Click Springs:

Consisting of a length of round steel rod with a handle at one end and a transverse slot at the other, bending tools are available from clock parts supply houses at a very nominal cost.

a.   Round Fixed Click Springs

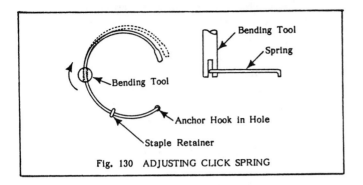

Fig. 130 ADJUSTING CLICK SPRING

If the click has not been removed from the wheel, disengage it from the slot in the click. Note the position of the end of the spring when it is fully relaxed.

As shown in Fig. 130, the slot of the bending tool is placed over the click spring at a point about one third of the way between the retainer and the end of the spring. To increase the pressure of the spring on the click, rotate the tool toward the end of the spring, applying only enough pressure to deflect the end slightly. Release and remove the tool and compare the location of the end of the spring with its former position. If it is still in the same place, more bending is necessary, so repeat the procedure, applying a little more pressure. Check again.

Reinsert the end of the spring in the slot in the click and check its action as outlined above. If the click was removed, put it and the pivot pin in position and check.

If the spring applies too much pressure, use the bending tool, in the opposite direction. Check again. Continue until correct action of the click results.

b. Flat Fixed Click Springs

In this case, the CLICK MUST BE REMOVED before adjusting. The pivot pin will be damaged during removal and a new one will be necessary.

Follow the same procedure as described for round springs. Temporarily install the pivot pin and click when checking spring and click action.

III. TIGHTENING LOOSE FLY (FAN) BLADES:

We have previously pointed out that fly blades should not be securely fixed to the arbor. On the other hand, they should not be so loose that they fail to rotate properly so as to apply controlled braking or governing of speed. There is only a friction connection between the blade and arbor, but there must be sufficient pressure to effect proper performance.

NOTE: To function properly, fly blades must be flat, both halves in the same plane. Frequently they have been bent out of plane in unsuccessful attempts to tighten them while they were still on the arbor. After removal from the arbor, it is a simple matter to restore flatness.

There are two basic types of fly blade connection to the arbor:

A. ONE PIECE FLY BLADES:

Made of thin sheet spring brass, one piece blades are formed to provide two slots with a narrow section in the center, between them. The outer sections of the blade have a groove to fit the arbor on one side. The narrow center section is grooved to fit the arbor on the opposite side. The center section is slightly bent to exert pressure on the arbor, to provide necessary friction.

To hold the blade in position, so it will not move along the arbor, a groove is provided in the arbor. In Fig. 131 you will see a narrow groove in the arbor into which a small dimple in the curved portion of the center section of the blade fits.

Another type employs a groove slightly wider than the width of the center section of the fly, which is then formed to fit that diameter. In both cases, the center section of the fly has been slightly bent so that it applies pressure to the arbor.

1. Tightening One Piece Fly Blades:

a. Remove the blade by sliding it toward the free end of the arbor. A little force will cause the center section to come free of the groove.

133

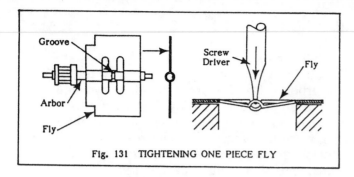
Fig. 131 TIGHTENING ONE PIECE FLY

b. Support the blade on vise jaws, with the curved part of the center section up, or on two blocks of wood or metal, so that the support is just at the ends of the slots forming the center section. Hold the ends of the blade firmly against the support with thumb and finger, then apply pressure with a small screw driver or other small tool, to the raised portion of the center section. Deflect it only enough to cause a little bending.

c. Slide the fly back on the arbor. Hold the arbor and flip one end of the blade sharply with a finger tip. It should spin only a few revolutions before coming to a stop.

If the blade spins freely, it is still too loose. Repeat the procedure, b, with a little more force. If it does not spin at all, it is too tight. Place the blade on the supports with the curved portion of the center section down and apply pressure as before, but in the opposite direction. Repeat until proper tension is achieved.

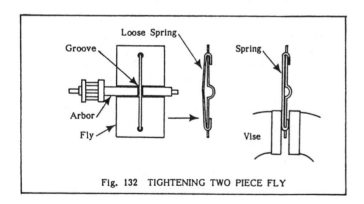
Fig. 132 TIGHTENING TWO PIECE FLY

B. TWO PIECE FLY BLADES:

In this case, the fly is again made from a thin sheet of brass. It has a transverse groove at the center to fit the arbor. Two small holes are punched at the

center, near the outer ends of the blade, into which a length of round steel spring wire is inserted from the grooved side of the blade, then bent over on the opposite side. This wire fits into a groove in the arbor and applies slight pressure to the arbor. The groove serves to hold the fly in position laterally on the arbor.

1. Tightening Two Piece Fly Blades:

a. Remove the blade by sliding it toward the free end of the arbor. If it is loose on the arbor, it will slide off easily.

b. Place one end in a vise, as shown in Fig. 132 and tighten the vise to squeeze one of the bent-over ends of the wire. Repeat the procedure for the other end of the wire. This should cause the wire to lie flush against the blade. Check to see if it does. If not, either the blade is not flat, or the spring is curved away from the blade.

c. If the spring is curved, place one end of the fly in a vise, so that the bent over end of the spring is just flush with the top of a jaw, then securely tighten the vise. Using a finger, apply pressure to the spring, just above the vise jaw and slightly bend it, toward the bent up end of the spring. Repeat with the other end of the fly.

d. Slide the fly into position on the arbor, so that the spring is in the groove. Flip one end of the fly sharply, with a finger. It should spin a few revolutions before stopping. If it spins freely the spring is not contacting the arbor in the groove and step c must be repeated.

IV. POLISHING PIVOTS:

You will recall that pivots are the small ends of arbors that fit into holes in the movement plates to form a bearing, so that the arbor can rotate freely, with minimum friction.

After disassembly, EVERY PIVOT should be inspected for wear or roughness of the surface, even though excessive wear of pivot holes was not evident.

Arbors are nearly always made from steel and their pivots machined and frequently polished, to provide a smooth surface. Inevitably, tiny particles of abrasive dust will work their way between the surfaces of the pivot and its hole, resulting in wear of both the pivot and the hole in which it runs.

Over a long period of time, this abrasion causes the smooth surfaces of both pivot and hole to become rough or uneven, and sometimes seriously grooved. Wear accelerates progressively to the point where the pivot is obviously very loose and the hole has been enlarged. These conditions eventually result in poor engagement of the teeth of wheels and pinions and, finally, in interference that prevents rotation and stops the movement.

Pivot holes enlarge as wear progresses and usually become oval in shape. When there is significant movement of the pivot in its hole, corrective measures should be taken. This will involve rebushing the hole, to provide a truly round hole of the proper size. We will discuss rebushing later.

You will have observed this condition in your movement, if it exists and if you followed the steps for examination of the assembled movement outline in Chapter 8.

A. WORN or ROUGH PIVOTS:

Regardless of whether or not you detected any significant wear of pivot holes in your first examination of the movement, after disassembly and cleaning, you should check each pivot carefully for roughness.

1. Examining for Wear:

Make sure that the pivot is free of dirt or any other surface contamination, then carefully inspect it, with a magnifying glass, if you have one, for irregularities such as pitting or grooving of the surface.

Another excellent check is to move the tip of a fingernail from the shoulder to the end of the pivot, as you rotate it. Any roughness will be felt. If the

surface is rough, or has visible grooves in it, smoothing and/or polishing will be necessary.

2. Smoothing and Polishing:

Smoothing involves removing metal from the pivot, down almost to the bottom of any pits or grooves, so that a nearly flaw-free surface results. Some experts say that pits up to ten percent of the pivot surface area are acceptable.

Polishing consists of bringing the surface to a high degree of smoothness.

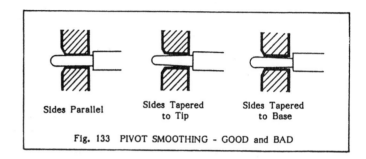

Fig. 133 PIVOT SMOOTHING - GOOD and BAD

a. Smoothing - Removing Pits and Grooves:

NOTE: Since smoothing will reduce the diameter of the pivot, it may be too loose in its original pivot hole. It may then be necessary to provide a new, smaller hole. Later in this Chapter we will show you how to perform this repair.

Since elimination of grooves requires the removal of a significant amount of metal from the pivot, it is important to maintain uniform diameter and roundness of the pivot as the work progresses.

If the sides of the pivot are not parallel, they will not bear evenly on the sides of the pivot hole and rapid wear will result. This is illustrated in Fig. 133.

The first illustration shows the sides of the pivot parallel, as they should be for good bearing. The second shows the sides tapered toward the tip, so that only a narrow section comes in contact with the pivot hole, near the base of the pivot. The third shows the opposite condition with the pivot tapered toward the base, providing bearing only at the tip of the pivot.

A clock maker would mount the arbor in a lathe and carefully cut the pivot to remove flaws. We have found that, for most clock work, satisfactory results can be obtained without the use of a lathe.

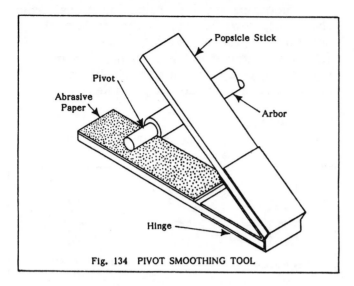

Fig. 134 PIVOT SMOOTHING TOOL

To help maintain parallelism, we suggest using a home made smoothing tool, consisting of two short hinged pieces of popsicle stick with a small piece of fine (No. 400) wet-or-dry abrasive paper cemented to one face. The pivot is placed against the abrasive and the top stick lowered to help keep it flat, then the arbor is rotated with the fingers, turning the pivot against the abrasive.

b. Making a Pivot Smoothing Tool:

(1) Obtain a popsicle stick, available at most supermarkets. With a fine saw, squarely cut one end, then cut the stick to a length of 5". mark the center and cut squarely. It is important that this cut be truly square.

An ideal tool for this purpose is a very small aluminum miter box and back saw which is available at hobby shops.

(2) Butt the ends of the two pieces of stick tightly together at the center cut. Holding both pieces firmly, place a piece of magic transparent tape (the kind you can write on) across the joint. Press down securely, then turn the assembly over and, with a sharp knife or razor blade, trim the edges flush with the sticks. Measure 3/4" on either side of the joint

and, with a sharp knife or razor blade, cut across the tape. Remove the outer ends of the tape.

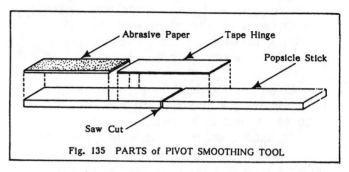

Fig. 135 PARTS of PIVOT SMOOTHING TOOL

(3) Cut a piece of No. 400 wet-or-dry abrasive paper about 1 1/4" long and a bit wider than the popsicle stick. Put a drop of white or yellow glue at the end of one of the sticks and smooth it to a thin film with the tip of a finger.

Locate one end of the piece of abrasive paper at the end of the stick and press firmly in place. Fold the two sticks together, making sure they are squarely mated, then clamp in a vise and allow to set for at least 30 minutes.

With the abrasive face down, use a sharp knife to trim its edges flush with the stick.

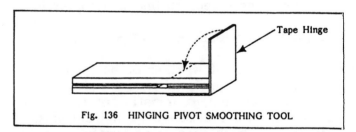

Fig. 136 HINGING PIVOT SMOOTHING TOOL

(4) Place the assembly in a vise with the hinged ends projecting. Adhere a piece of tape to the face of one stick and firmly press it in place, then fold it over the ends of the two sticks and press firmly (Fig. 93). Pull the tape over the second stick and press into place. Trim the ends and sides of the tape. The smoothing tool is now ready to use.

c. Using the Pivot Smoothing Tool:

With the arbor of the pivot to be smoothed between the thumb and forefinger of one hand, and the smoothing tool held by the fingers of the other, place the pivot on the abrasive face of the tool.

Bring the two faces of the tool together, applying slight pressure with thumb and forefinger right over the pivot. This will grip the pivot in full contact with the abrasive paper.

Keeping the arbor square with the tool, use the thumb and forefinger of the other hand to gently rotate the arbor, back and forth against the abrasive paper.

If both the tool and the arbor are lightly gripped, you will be able to feel that the pivot is properly located. By holding the open end of the tool so you can see into it, you can easily visually align the arbor with the faces of the tool.

Inspect the pivot frequently to see how the work is progressing and stop as soon as flaws are removed.

d. Polishing:

We recommend making another tool for polishing, identical to the smoothing tool, except that a piece of crocus cloth is used instead of the wet-or-dry abrasive paper.

Crocus cloth is a very fine polishing medium and is available from automotive supply outlets. Use this second tool in the same way as the smoothing tool. It will provide a highly polished pivot surface.

B. HARDENED PIVOTS - BREAKAGE:

In better quality movements, the pivots are heat-treated to harden them and are then highly polished. Hardened and polished pivots are much more resistant to wear than ordinary ones. They are usually smaller in diameter and much more brittle than those that are not hardened.

Hardened pivots are likely to break off if any attempt is made to bend them, or if they are subjected to sharp stress, as when a spring breaks, or if excessive force is used in positioning them during assembly of the movement.

Replacing a broken pivot requires the use of a precision lathe and we suggest that you have this work done by someone who has the necessary equipment.

V. BUSHING WORN PIVOT HOLES:

When a pivot hole becomes worn to the point that it affects the performance of the movement, or if a pivot has been reduced in diameter during smoothing, it becomes necessary to provide a new hole.

This is done by creating a larger hole in the clock plate into which is inserted a tightly fitting bushing with a hole of the proper size to correctly fit the pivot.

Clock makers who do lot of bushing work use rather expensive tools designed specifically to perform this task, or utilize their lathes and other machines. One advantage of these methods is that, when properly used, they insure that holes for the bushings are precisely located and absolutely perpendicular to the plate. These considerations are important, particularly when dealing with very small pivots in comparatively thick plates, as in high quality French movements.

For the more common movements, such as most of those of American manufacture, very satisfactory results can be obtained without such equipment.

One method makes it possible to install new bushings without disassembling the movement. We do not recommend this system because it is impossible to correct pivot problems which may be the cause of hole wear and because it creates tiny brass chips which will find their way into the movement and cause trouble. The equipment required is comparatively expensive and the end result leaves much to be desired.

A. HAND BUSHING SYSTEM:

There is a simple method of bushing, requiring only hand tools that we have found quite satisfactory. It will require a modest investment of around $25.00 for the tools and about $12.00 for a basic

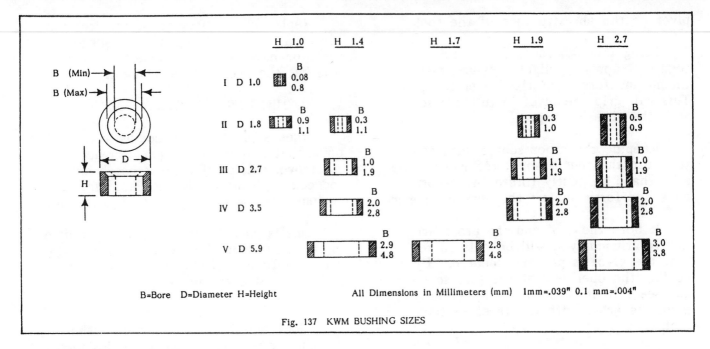

		H 1.0	H 1.4	H 1.7	H 1.9	H 2.7
I	D 1.0	B 0.08 0.8				
II	D 1.8	B 0.9 1.1	B 0.3 1.1		B 0.3 1.0	B 0.5 0.9
III	D 2.7		B 1.0 1.9		B 1.1 1.9	B 1.0 1.9
IV	D 3.5		B 2.0 2.8		B 2.0 2.8	B 2.0 2.8
V	D 5.9		B 2.9 4.8	B 2.8 4.8		B 3.0 3.8

B (Min)
B (Max)
D
H

B=Bore D=Diameter H=Height   All Dimensions in Millimeters (mm)   1mm=.039" 0.1 mm=.004"

Fig. 137 KWM BUSHING SIZES

assortment of bushings. When compared with the cost of having even one clock rebushed by a professional, this is a very reasonable investment. It uses standard KWM bushings which are readily available from clock parts suppliers.

B. KWM BUSHINGS:

Bushings are normally described in terms of their height (H), diameter (D) and bore (B) or hole size. Suppliers' catalogs list available sizes numerically, by the part number used to order specific individual bushings.

To give you a better idea of the size combinations, we have developed Fig. 137 to help you visualize what you might need.

At the left of Fig. 137, a typical bushing is shown in plan and cross section. The letters H, B and D correlate with the catalog listings and the chart to the right in this figure.

1. Selecting a Bushing:

a. Bushing Height (H):

The height of a bushing should be at least as great as the thickness of the plate into which it will be inserted. This is the first consideration in selecting a bushing. Most American clocks have plates close to 1.4 or 1.9 mm. See the chart, under H 1.4 and H 1.9. Within these heights, there is a broad selection of bores to fit a wide range of pivot sizes.

b. Bushing Bore (B), or Hole Size:

The second consideration in selecting a bushing is the size of the pivot hole. This can only be determined after the pivot has been smoothed and polished. The simplest method is to have an assortment of bushings in the range of heights and hole sizes needed and to select one that looks close and trying it on the pivot.

Work from larger to smaller bore sizes until you come to one that will not fit over the pivot. The next larger size will probably be right. Since bushings are quite small and difficult to handle, we suggest using a pair of tweezers for holding them.

On the chart in Fig. 137, the range of available bore sizes are shown to the right of the bushings illustrated. Note that there is a range of bore sizes for each diameter. The diameters of the bushings increase progressively as the bore size increases.

In the sizes used for most clocks, bores will range from 1.0 mm upward in increments of 0.1 mm (.004"). Thus, if your assortment progresses in 0.1 mm bore sizes, when you come to a bore that won't go over the pivot, and the next size larger does, you know that the pivot diameter must be more than the smaller and less than the larger. For our purposes, we can assume it is closer to the larger diameter and be right within 0.005 mm, or .002".

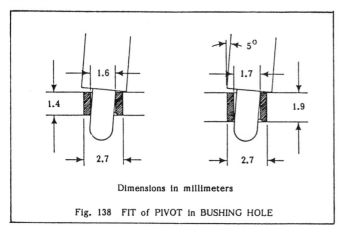

Dimensions in millimeters

Fig. 138 FIT of PIVOT in BUSHING HOLE

1). Fit of Pivot in Bore Hole:

Contrary to what you might think, pivots should fit rather loosely in pivot holes. Most authorities agree that there should be sufficient clearance to allow the arbor to rest at an angle of 5 degrees.

In Figure 138, we have illustrated how the clearance between pivot and hole is affected by the length of the hole, resulting from an increase in the height of the bushing.

We have assumed a pivot diameter of 1.5 mm. With a bushing 1.4 mm high, a bore (hole diameter) of 1.6 mm allows the arbor to tilt 5 degrees. When the bushing height is increased to 1.9 mm, the bore must be increased, by 0.1 mm, to 1.7 mm.

It is much better to a have a little more than necessary clearance between pivot and hole, than to take a chance on having too little.

For your first bushing effort, we recommend that you select a bushing 0.2 mm larger than the nominal size of pivot as determined by trying bushings, as outlined above.

c. KWM Bushing Assortments:

KWM bushings are available in assortments of sizes, usually containing ten each of ten different bushings, all of the same height and usually of two diameters. Bores are progressive, in 1 mm or 2 mm increments. Currently, most assortments are priced at under $13.00.

Plates of most earlier American clocks are about 1.4 mm thick, so that is the height of bushing required. Common alarm clocks also use this height. Later clocks have plates close to 1.9 mm thick, so that height bushing should be used.

Figure 137 shows that there is a wide range of bore diameters available in both these heights. We have found that the most commonly needed bores are those between 1.3 mm and 2.8 mm. A single assortment in this range is available in heights of both 1.4 mm and 1.9 mm.

Each assortment is packaged in a plastic box with ten compartments, each containing one size of bushing. The cover of the box has a chart with boxes corresponding to these compartments, showing for each compartment the full identification of its contents. The following is an example of one of them:

Identification	Meaning
H 1.9	The Height is 1.9 mm
B 1.5	The Bore is 1.5 mm
D 2.7	The Diameter is 2.7 mm
L 42	The number identifying this bushing
III	The Broach needed to make the hole

Without a micrometer, it is almost impossible to determine the diameter of a pivot, except by the go-no-go method we have described. Plate thickness, however, can be measured closely enough with a good centimeter rule to determine whether it is near to 1.4 mm or 1.9 mm.

With half millimeter markings on the rule, the first will be obviously close to the 1.5 mm mark and the second close to the 2.0 millimeter mark. The appropriate assortment can then be ordered.

B. HOLE SIZES for BUSHINGS:

To insert a bushing in the plate of the movement, the pivot hole must be enlarged. This is accomplished through the use of cutting broaches, or reamers, accurately ground to produce a hole of exactly the right size to provide a tight fit when the bushing is pressed into it.

1. Bushing Diameters:

In the left hand column of the chart in Figure 137 is listed the diameter, D, of all bushings shown to the right. Each diameter is also identified with a Roman Numeral, indicating the broach to be used to make the hole for that diameter of bushing.

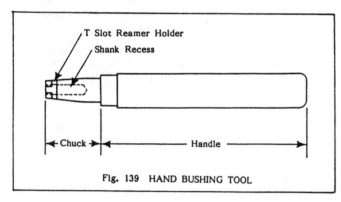

Fig. 139 HAND BUSHING TOOL

2. Hand Bushing Tool:

Illustrated here is a simple tool, consisting of a handle with a T-slotted steel chuck, into which any of 5 cutting broaches (reamers) and a chamfer cutter may be inserted. It currently is priced at under $25.00, complete with a set of reamers and a chamfer cutter.

The shank of the reamers fits snugly into hole in the chuck and a pin on the shank engages in a T-Slot in the chuck, so that force can be applied in either direction.

a. Identifying Cutting Reamers:

The reamers are of progressively larger diameter, each matching a bushing diameter. Fig. 137 identifies each size by a Roman Numeral, I through V. Each is made to provide the proper size of hole for a particular bushing.

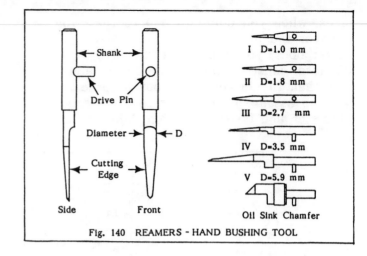

Fig. 140 REAMERS - HAND BUSHING TOOL

In Figure 140, we have shown that diameter, D, for each reamer. The actual size of the hole it will cut, however, is slightly smaller than the diameter of the bushing for which it is intended. This is to provide a tight friction fit that requires forcing the bushing into the hole, thus insuring that it will stay in place.

b. Reamer Construction:

Each reamer is made of hardened tool steel. The shank fits into the chuck of the handle and carries the drive pin. The portion immediately below the shank, D, is ground to the exact diameter of the finished hole. Below this straight section the broach tapers to a point and is ground flat to provide a sharp cutting edge on each side. It will cut when turned in either direction.

It should be noted that these reamers are designed for hand use and should never be used in powered equipment, such as an electric drill. They may be used in a drill press, only if the chuck is turned by hand, never under power.

3. Using the Reamer to Cut Bushing Holes:

a. Cutting Action

Unlike a drill, which cuts only at its tip, a reamer is designed to cut on its sides. Thus, a starting hole is necessary. Reamers, then, are used to enlarge holes and the process of doing so is called "reaming".

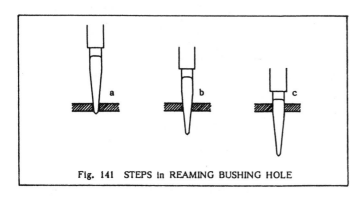

Fig. 141 STEPS in REAMING BUSHING HOLE

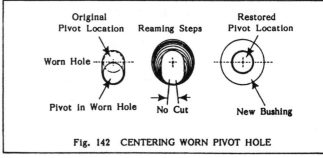

Fig. 142 CENTERING WORN PIVOT HOLE

Since the tapered point carrying the cutting edges gradually increases in diameter, the hole it enlarges is also tapered, until the maximum diameter is reached. Fig. 141a shows the beginning of broaching, with the tip just entering the starting hole, in our case the worn pivot hole.

Fig. 141b shows the hole partially enlarged and Fig. 141c, the hole enlarged to its final diameter with the straight section completely through the plate.

b. Important Cautions in Using a Reamer:

(1) Keep the reamer at 90 degrees to the face of the plate at all times during cutting. Be especially careful to maintain this angle as you approach the straight side of the cutting area and maintain it carefully until the straight cutting section of the reamer is through the plate. If care is not exercised, the final hole may be over size and the bushing loose, or it can be at an angle that will not permit the pivot to enter properly.

(2) Very little force is necessary when cutting with these reamers. Exert only gentle pressure on the reamer while cutting. Heavy pressure will make the process faster, but can have dire consequences. Take time to be careful and safe. We find that placing a finger tip on the end of the handle and rotating it between thumb and forefinger is the best method to use.

4. Centering the Bushing:

The major objection voiced against the hand bushing method is that the center of the bushed hole cannot be restored to that

of the original pivot hole. Professional bushing tools are designed so that they rigidly position the plate and the reamer to insure that the point of the reamer is at the exact center of the original pivot hole and is firmly held there during cutting.

We have found that, using reasonable care with the hand method, the pivot hole in the new bushing can be brought very close to the original pivot hole position.

Bushing is usually required when the pivot hole has been elongated to an oval shape by about half the diameter of the pivot. Based on this premise, we will illustrate the steps involved in making a new hole with its center closely matching that of the original, as shown in Fig. 142.

a. Centering Procedure:

(1) Locate Position of Original Pivot Hole:

Wear of pivot holes is in the direction of force applied to the pivot and results in elongation of the hole to an oval shape. While this wear is taking place, the pivot works on the worn end of the hole, as shown in Figure 142. Before disassembly, it is easy to verify this by observing its location in the hole while force is applied by the spring or weight, or by manually rotating the wheels in the direction of normal operation.

The end of the worn hole opposite to that in which the pivot works is part of the original pivot hole.

The new hole for the bushing will be significantly larger than the original pivot hole, to accommodate the necessary wall thickness of the bushing. In effect, we will use the bottom end of the worn hole as the limit of the diameter of the new, larger hole. This will locate the center of the new pivot hole very close to that of the original, as demonstrated by the drawing of the new bushing in Fig. 142.

(2) Define the Center of the Original Pivot Hole:

Half of the original pivot hole will remain at the unworn end of the existing hole. Eyeball this end and estimate the center of the original hole. Using a straight edge and the tip of sharply pointed knife, or razor, lightly scribe a line, passing through the original center, on the plate on either side of the hole about 1/8" long. This is indicated by the dotted line in Figure 142.

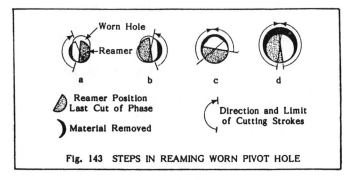

Fig. 143 STEPS IN REAMING WORN PIVOT HOLE

(3) Do Not Cut the Worn End of the Hole:

As shown in Figure 143, keeping the smooth side of the reamer at the worn end of the hole, then alternately cutting left and right, will maintain the worn end, while enlarging the hole in the direction of the original pivot hole.

(4) Fig. 143a shows the start of the reaming operation, with the round side of the reamer at the lower right of the worn portion of the hole. A clockwise cutting stroke removes material at the lower left side. Note that the cutting stroke stops when the cutting edge reaches the edge of the worn hole.

(5) In Figure 143b, the procedure is reversed and material is removed from the lower right side of the hole.

(6) As cutting continues, there will come a time when the reamer begins to cut above

the original hole, Figure 143c. It is important to note, however, that the broach is tapered and the cutting progressive as it goes deeper into the plate.

(7) Continue taking alternate clockwise and counter clockwise strokes until:

(a) The circle being cut by the reamer, as seen on the top plate, is centered on the line you scribed on the plate or

(b) The end of the tapered section of the reamer is nearing the top surface of the plate.

When either of these conditions occur, finish by turning the reamer continuously in one direction until the cutting edge of the straight side of the reamer barely passes through the back face of the plate. This is shown in Fig. 141c. The hole will be of the desired diameter and the sides will be straight and parallel.

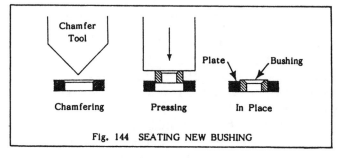

Fig. 144 SEATING NEW BUSHING

5. Seating the Bushing:

Reaming the hole will probably develop a slight burr at the top of the hole. Use the chamfer tool, Figure 140, with only a light revolution or two, to barely remove this burr and provide a tiny chamfer to assist in centering the bushing.

With the back of the plate on a firm surface, locate the bushing, with its chamfered end up, in the hole. Squarely center a short length of metal rod with a flat face on the bushing, as shown, then, keeping the rod perpendicular, gently tap the bushing into place. When fully seated, its bottom end should be flush with the back surface of the plate.

Check to make sure the bushing is firmly held in place by inserting the end

142

of a round toothpick in the hole, applying a little pressure and trying to rotate the bushing. It should not turn, or be moved beyond the back face of the plate.

If everything has been done as it should, you will have a firmly seated bushing, properly located, with a hole of the correct size for the pivot.

If the bushing is not firmly held in place by friction it must somehow be secured. There is an undesirable, but workable solution to this problem. It must be executed with great care.

Remove the bushing from the hole, then press a round tooth pick firmly into the pivot hole and rotate it a few times. This will slightly compress the part of the pick in the hole and leave a visible shoulder on it at the top of the bushing.

Remove the toothpick and cut the tip back so that its length from the shoulder created by the bushing is a tiny bit less than the height of the bushing. Reinsert it firmly in the pivot hole of the bushing.

Place the movement plate, back down, on a piece of waxed paper laid on a smooth surface. Barely wet the tip of another tooth pick with instant (cyanoacrylate) glue and quickly rotate the outside of the bushing against it. Be very careful to get no glue near the pivot hole. Insert the bushing in the prepared hole and press to the back of the plate. Carefully release the tooth pick and leave it in place in the hole of the bushing for at least ten minutes. Turn the tooth pick a few times to make sure the bushing is secure, then remove it.

SUMMARY

Now that you have gained a basic understanding of how pendulum clocks work, some of the things that can make them stop operating and how to correct them, we hope your interest in clocks will continue to grow.

INDEX

145

BIBLIOGRAPHY

SELECTED SUPPLEMENTAL REFERENCE BOOKS

Balcomb, Philip E.   The Clock Repair FIRST READER, Second Steps for the Beginner
                                Tempus Press, 1991
Britten, F. J.   Watch & Clock Maker's Handbook, 11th Edition
                                Baron Publishing Ltd., 1976
Bruton, Eric   Dictionary of Clocks and Watches
                                Bonanza Books, 1963
de Carle, Donald   Practical Clock Repairing, 2nd Ed.
                                N.A.G Press Ltd., 1968
Harris, H. G.   Handbook of Watch and Clock Repairs
                                Emerson Books, Inc., 1974
Goodrich, Ward L.   The Modern Clock, Eighth Printing
                                North American Watch & Tool Supply Co., 1968
Jones, Bernard E.   Clock Cleaning & Repairing, 2nd Ed.
                                Cassell Ltd., 1954
Kelly, Harold C.   Clock Repairing as a Hobby
                                Association Press, 1972
Penman, Laurie   The Clock Repairer's Handbook
                                Arco Publishing, Inc. 1985
Plewes, John   Repairing and Restoring Pendulum Clocks
                                Sterling Publishing Co., Inc. 1984
Smith, Eric P.   Repairing Antique Clocks
                                Redwood Burn Limited, 1973

Check your local public library and book sellers for these and other titles relating to clock repair. If they cannot help you, a good selection of books on clocks are available from the the following mail order dealers who specialize in books on horological subjects:

Adams Brown Co.   266 N. Main Street, PO Box 357, Cranbury, NJ 08512
Arlington Book Co.   PO Box 327, Arlington, VA 22210-0327
Meadows & Passmore   Farningham Road, Crowborough, E. Sussex TN6 2JP, England
Scanlon Horological Books 112 Ulrich Ave., Modesto, CA 95353

Many Museums stock books on Clocks for sale to visitors, but may not sell by mail.

Clock Parts Supply houses, such as those listed on page 151 are a very good source of horological books.

BOOKSELLERS SPECIALIZING IN HOROLOGICAL BOOKS
(Revised 12/20/1994)

While this list is probably not complete, it includes all Booksellers known to the Author as specializing in books related to timekeeping, at the time of publication of this edition. In addition, most horological museums known to carry books on the subject are listed.

Most dealers in Clock Repair Parts and Supplies also stock books relating to clock repair. -See next page.

Adams Brown Co.   26 North Main, P.O. Box 357, Cranbury, NJ 08512

ALL ARTS BOOK SHOP   160 Oxford Street, Woollahra, 2825 AUSTRALIA

Arlington Book Co.   P.O. Box 327, Arlington, VA 22210-0327

Meadows & Passmore Farningham Rd., Crowborough, E. Sussex, TN6 2JP, ENGLAND

Charles M. Murray PO Box 4518, Postal Station D, Hamilton, Ontario L8V 4S7, CANADA

NAWCC Museum   514 Poplar Street, Columbia, PA 17512

Rita Shenton, Horological Bookseller   148 Percy Road, Twickenham TW2 6JG, ENGLAND

Scanlon American Reprints   1112 Ulrich Avenue, Modesto, CA 95350

The TIME MUSEUM   7801 East State Street, Rockford, IL 61125-0285

CLOCK PARTS SUPPLIERS
(Revised 12/20/1994)

This is by no means intended as a comprehensive listing of all firms who engage in the selling of clock parts and supplies. The firms shown are known to the Author and carry a reasonably comprehensive stock of supplies. Most also issue comprehensive catalogs, for which there may be a nominal charge, usually refunded as a credit with your first order.

Empire Clocks	1295 Rice Street, St. Paul, MN 55117-4591
S. LaRose, Inc.	234 Commerce Street, Greensboro, NC 27420
Murray Clock Craft Ltd.	510 McNichol Ave., Willowdale, Ontario, M2H 2EI, CANADA
Merritt's Antiques	R. D. 2, Douglasville, PA 19518-0277
PRIMEX Inc. (ClocKit)	1500 N. Elkhorn Road, Lake Geneva, WI 53147
R & M Imports	3313 Harlan Carroll Road, Waynesville, OH 45068
Ronell Clock Company	P.O. Box 5622, Grants Pass, OR 97527
Southwest Clock Supply, Inc.	335 South Main St., Carthage, MO 64836
TIMESAVERS	16939 E. Colony Dr., Suite 2, PO Box 18339, Fountain Hills AZ 85268
Louis Williams Clock Supplies	755 Woodbine Blvd., Jackson, MI 49203
Woodcraft Supply Corp.*	7845 Emmerson Ave., P.O. Box 1686, Parkersburg, WV 26102

* Retail Stores in many major cities. See Telephone Directory.

There are other firms that specialize in one product, or a limited line of products, as well as other general line parts houses. Many advertise in the NAWCC MART, which is sent every other month to members of the National Association of Watch and Clock Collectors.

If you live in a larger city, you may want to check the yellow pages of your local telephone directory under "Clock Supplies" for possible local sources.